Andrew Davies

Mapping the Mountains

Sella Margedi
C. Medio Nöbande S.
C. Alto Nobande
Piede Sella Biacerahi
Biacerahi
Ghiacciaio Nöbande Söbande
Ghiacciaio Choktö
C. Choktö
C. Lamidukpa
Lamidukpa
Chari-Chor
Chongulter
Shushing
Ghº Dumultar
Dumultar
C. Dumultar

Andrew Davies

Mapping the Mountains

Foreword by

Alastair Bonnett

Dutch
Mountain Film
Festival

Glacier de la Plate
des Agneaux
Pic de Neige
Tête de Charrière
Roche d'Alvau
Sommet de Roche Faurio
Glacier
Blanc
Glacier
Blanc
Glacier
de
l'Encula
la Maye
Cabane des Bergers
la Bérarde
Glacier de la Bonne Pierre
Pic de Bérarde
Crête de l'Encula
les Ecrins
Glacier du Vallon
Glacier de la Temple
Glr. de la Coste Rouge
Glacier
Noir
Mont Pelvoux
Cabane des Bergers
Rochers des Boeufs Rouges
Tête de Chéret
St. de l'Aile Froide
Coste de Sialouze
Gd. Glacier
de
la Pilatte
Glacier du Sélé
Col Says
Montre du Says
Giberney Mgne
les Bans Mt.
Crête des Boeufs Rouges
Pic des Opillous

A Sense of Wonder

Maps of mountains are beautiful, but they can also be wild and capricious. It is no easy matter to capture vertiginous rock, tumbling scree, and crashing waterfalls, from above, and for the resulting map to be both accurate and useful; even a vital tool that can guide people to safety. The story of mountain mapping is a drama, full of difficult journeys and knotty puzzles. Andrew Davies has brought this drama alive for us, and each page of Mapping the Mountains speaks of the effort, ingenuity, and audacity behind our attempts to cartographically conquer the world's summits.

It is a long history. The earliest maps in this book show that mountains were not always the objects of fascination that they later became. Ancient map makers were mostly interested in linking and locating human settlements. Mountains loom across their maps like dark shadows, places beyond the ken of civilisation. It is only with the emergence of modern techniques, Enlightenment hutzpah and romantic wonder that the lure of mountains took flight. Many of the stories gathered here were new to me and they are all extraordinary. Who cannot admire the dogged ambition behind the travails and travels of the group of cartographers sent by French Academy of Sciences in 1735 to measure the size and shape of the Earth. Setting up a triangulation station on Pichincha, a volcano high above the city of Quito, they were, as Andrew Davies tells us, 'pummelled by snow and hail, spent much of their time huddled together trying to keep warm and in the mornings had to be dug out by their local porters from the snow drifts'. Two and half years after leaving France 'they had completed their first triangle and there were another 30 to go'.

Today it is easy to take maps for granted. We all have instant and incredibly accurate maps on our phones. Our sense of wonder has been deadened. It needs to be rekindled. Mapping the world's heights was an immense and often heroic task. This is a book of science but also adventure. It takes us across the Alps and the ranges of Europe, to the Americas and Asia, as well as to Africa and Ethiopia's Rwenzori mountains, where explorers faced freezing heights and swampy mires and camped in rain and fog; 'Every time that anyone stepped out of the camp', one diary entry from the Rwenzori team found by Davies notes, 'he would sink into the mud'.

The lure of mountains is not fading. I suspect it is getting more powerful. Our crowded, industrial civilisation is getting more crowded and more industrial. Cast upon any map our eye is drawn to those places where contour lines gather in dark, exciting, clots; where there are no shopping malls or police sirens. Mountains are forbidding yet we yearn for them. Turn the pages of this book and let your imagination soar.

Alastair Bonnett

Contents

East Rongbuk Glacier
Camp II
22580
Khart
20100
20000
21000
Camp III
Changtse
24730
Camp IV
Chang La
19000
19400
Camp V
Camp VI
23660
Mt EVEREST
29002
25700
Lhotse
27890
18000
17690

Contents

Worldmap

10 Napoleon's triangles
page 110
7 William Roy's Great Map
page 76
8 The stargazer
page 88
5 The Cassinis and the birth of modern mapping - page 60
18 Mountaineer turned cartographer - page 204
9 The unlikely mountaineer - page 98
13 Twenty-one summers - page 138
Swiss-style
3 On the itinerary
page 42
6 Oranges and lemons
page 70

Main locations

Introduction

Mapping the Mountains

map *noun*: a diagram of any part of the Earth's surface, showing geographical and other features
mountain *noun*: a very high, steep hill, often one of bare rock

Chambers 21st Century Dictionary

The film *The Summit: A Chronicle Of Stones to Serenity* weaves a spellbinding tale of adventure and intrigue which pits a team of military surveyors against a group of gentlemen alpinists in a race to climb an untrodden and seemingly impregnable peak in the Japanese Alps. It is 1907 and Shibasaki Yoshitaro, the military engineer charged with mapping the rugged mountain terrain, has been issued with strict orders to reach the summit of Tsurugidake before the amateur mountaineers. Anything less would besmirch the honour of the Japanese Imperial Army. Risking life and limb to overcome snowstorms, avalanches and rockfalls, as well as an endemic belief that the mountains are the sole domain of the gods, Shibasaki and his team of surveyors and local guides, encumbered by hefty surveying equipment, finally arrive at the top to claim victory, only to discover a sword and a priest's staff on the summit that had been left there by ascetics a thousand years previously. *Tsurugidake: ten no ki*[1], published as a book in 1977 and later released as a highly acclaimed film adaptation in 2009, is a fictionalised account, yet nevertheless based largely on real events which have been embellished to dramatise the storyline. Whilst the plot may not be wholly factual[2], the outcome is a rollicking yarn, wonderfully evocative of the heydays of mountain exploration, when a cocktail of scientific endeavour, adventure, espionage, perseverance, rivalry, comradeship, heroism, survival and sometimes tragedy came together. Improbably, as we will discover on this odyssey of mountain mapping, real-life cartography too, is full of larger-than-life tales.

* * *

My own journey starts on an ancient hillfort in North Wales. The limestone outcrop of Bryn Euryn, not much more than 100 metres in height, rises above the small seaside settlement of Rhos-on-Sea. Its summit was once the stronghold of Cynlas Goch, a Welsh king who held sway in these parts in the early sixth century. Climbing the hill in the summer holidays was considered a rite of passage in our family, and when I was old enough, I clambered up with my mum, dad and older brothers. On reaching the top, it wasn't the Dark Age fortress that left me spellbound, but instead the concrete pillar around four feet in height, standing on Bryn Euryn's summit. The exhilaration of being held aloft by my father and placed on top of the column remains with me to this day: it was my first-ever encounter with a trig point and it was love at first sight.

Anyone who roams the open spaces of the British mainland will be familiar with these triangulation pillars – popularly known as trig points – with over 6,000 of them adorning prominent elevations around the country. They were installed by the Ordnance Survey in the early twentieth century as part of a massive re-triangulation network to map the country more accurately. As I grew up and scaled ever higher hills, these concrete pillars came to represent the culmination of a long and often arduous hike, at all times accompanied by a detailed OS map.

Around the same time as this first encounter, my father – who worked as an insurance clerk in Manchester – came home one evening with a brown-paper parcel containing a framed picture which I still have today. It depicts a snow-covered mountain, its almost perfect pyramidal shape rising impossibly above the lush green meadows below. In the foreground is a line of four timber haylofts, which drop away centre right and in the background, against an otherwise brilliant blue sky, are faint wisps of cloud around its summit. That mountain – the Matterhorn – clearly cast a spell on him. He developed a curiosity for the mountain which likewise made its mark on me. Our local library in Chorlton-cum-Hardy held a copy of Edward Whymper's *Scrambles Amongst the Alps*, a title that was otherwise long out of print, so in order to satiate our interest in the author's escapades on the mountain we had to keep renewing the loan on the book. In my early teens, '*Scrambles*' was an unputdownable read.

Over half a century and many trig points later, I was in a pub enjoying some beers when a proposition was put to me. Would I be willing to curate an exhibition on maps for the Dutch Mountain Film Festival? My drinking partners that evening, the organisers of the festival, insisted that I would be the ideal person, with a huge collection of topographic maps, a penchant for tales of adventure, a passion for the outdoors and, perhaps not unimportantly, as a professional translator, I was handy with words. So, with some Dutch courage inside me (probably the intention), my resistance inevitably buckled. The exhibition *Mapping the Mountains* in 2019 proved a success and – when it became clear we'd only touched the tip of the iceberg – I was asked to write the book. This time, there was no need for hesitation (or intoxication), even though it has been a hard road and several years in the making.

* * *

What makes a mountain map inherently different from any other kind of topographic map? After all, the creation of any map depicting landforms is based on universal cartographic principles: the measurement of angles and distances. However, where a mountain map fundamentally differs from other, say lowland maps, is that it incorporates the addition of a third dimension in its representation of topography. In mountainous terrain, the cartographer has to overcome two key obstacles. The first relates to having to work in uneven, poorly navigable terrain to measure angles and distances which are vertical in nature and often where the field of vision is obstructed by intervening terrain. The second is to translate this vertical dimension onto a flat surface (i.e. the map) in a way and at a scale which is intelligible to the reader of that medium.

Nowadays, topographic maps have reached unprecedented levels of accuracy and detail. This technical brilliance is matched only by their beauty. How is it possible, you may ask, to transform a tangled mass of mountains into two dimensions and cause them to come away from the page in a trick of the eye? Of course, this near-perfection – can a map ever be perfect? – has not been achieved overnight, but over centuries of evolving cartographic skills. The carefully crafted lines, hatching, contours, colours and shading are techniques that have been applied and refined by generations of cartographers to produce this illusion, based on painstaking measurements of heights, distances and angles taken in the most inhospitable of environments.

If there is a story of mountain cartography to be told, we should first recognise that no joined-up narrative exists. Literature on cartography and on mountaineering abounds, but seldom in combination and only then limited in geographical or historical scope, restricted to specific expeditions or campaigns. Amongst this vast body of literature however, there is a thread by which to construct a timeline. Throughout history, funding for exploration – state-sponsored or otherwise – had to be justified by meticulous documentation and scientific content, so accounts of expeditions to 'undiscovered' lands drew heavily not only on acts of human endeavour, but also on their scientific results, of which cartography was frequently part. As we will see, the development of mountain cartography is inseparably bound up with the history of exploration and discovery. In the case of mountain exploration, for example, mapping was considered crucial to the successful execution of a campaign: when the Mount Everest Committee was set up in 1920 with the intention of putting a British subject on its summit, they first sent out a reconnaissance mission with the specific aim of mapping the mountain and identifying assault routes. A craving for unadulterated adventure therefore, was balanced with a quest for greater scientific understanding, but it is sometimes difficult to establish which was subordinate to the other.

The overlap means that this book is not a self-contained history of mountain cartography, but rather a collection of historical essays which give a snapshot of how mapping in the mountains evolved over time. The stories recall actual expeditions or missions, across the globe, which were groundbreaking in their own right *and* which ultimately resulted in a map. In the main, the leading players are either mountaineers or cartographers and, as in the case of Bradford Washburn or Paul Helbronner, sometimes a mix of the two. Most, but not all these stories are bookended into a period lasting roughly from the mid-17th century, when technical advances gave rise to scientific discovery, up until the mid-20th century, a time which ushered in cartographic practices such as aerial reconnaissance and later, automation and computerisation. By the millennium, tedious fieldwork had thus been rendered largely redundant.

To get the most out of these stories, a basic understanding of the principles on which topographic mapping is founded and of the instrumentation surveyors used in the field is required. Once data had been collected in the field this had to be transformed by office-based cartographers into a format readable on paper, the map itself, and to attain this legibility, relief depiction was a skill that mapmakers had to master. As a science, cartography advanced on the back of major breakthroughs in the application of these principles and improvements in instrumentation. Over the centuries, mapping was influenced by a growing thirst for knowledge, technological innovation, economic development, nationalism, military antagonism and more latterly, sport and recreation, often working in tandem with each other. These key advances, where relevant to mountain cartography, are included in the book as historical highlights.

There are two caveats when reading this book. First, it is important to bear in mind that the book adopts an essentially Eurocentric perspective, especially in the main period covered by the essays. It is often said that maps are made by the victors: the predominant worldview was driven by the colonising European superpowers who wanted to show the world as they interpreted it. Evidence of any pre-colonial mapping was tacitly pushed into the shadows through cultural appropriation. It might be assumed that mountain maps, which show topography rather than political divisions, would be immune from this process, but this is inherently not the case. Riding roughshod over indigenous place names, for example, has been the modus operandi of occupiers since time immemorial and this was no less prevalent in the naming of mountain features, Mount Everest being perhaps the mother of all examples. Russia, and later the Soviet Union, adopted this practice in the Pamir. The Canadian Rockies has an outlandishly random British-leaning nomenclature introduced around the time of their initial tourist and economic development. Likewise, German-speaking South Tyrol underwent a comprehensive Italianisation of names when the region was ceded to Italy after the First World War.

Second, apart from this book's passing references made to Fanny Bullock Workman (in *Off the Map*) and Barbara (*née* Polk), the wife of Bradford Washburn, one stark omission is the role played by women in cartography and (mountain) exploration. The concept of gender equality barely existed in Western cultures before the end of the 19th century, so these were domains that were almost exclusively reserved for men. Sons, rather than daughters, pursued professions practised by their fathers, in what seemed like a perpetual time-honoured cycle. Original works documenting the achievements of mould-breaking women in the

Great Outdoors have been frustratingly off the radar too, though in recent years the number of publications addressing this oversight has happily been increasing. As far as mapping is concerned, the legacies of 19th-century women explorers like Mary Schäffer Warren and Mary Vaux in the Canadian Rockies, for example, are worthy of further investigation.

* * *

The techniques first devised by ancient civilisations to make navigation easier through the observation of celestial objects and the measurement of angles gradually became translated into the form of a map. Over time, these techniques evolved and became ever more sophisticated, ushering in increasingly accurate maps and safer navigation. Mountain mapping was no different in this respect and was born out of a necessity to make sense of an extreme form of three-dimensional topography. This need only intensified when mountaineering as a sport developed and tourism started to open up mountain regions to the masses. In more recent history, it has been recreation which has become the major catalyst to improving these maps, helped along by technological advances that have dispensed with the need for on-the-ground teams of surveyors and office-based draughtsmen. Artificial intelligence (AI) is the latest trend which is making an impact on mapping in general.

In our modern world, on the back of these technological advances, we have reached a point where getting lost is almost impossible and ironically, we have now amassed so much data that maps themselves, in their conventional form, are no longer necessary for us to move around from place to place. Our mobile devices allow us to navigate and fix our location with pinpoint accuracy. Getting from point A to point B, even in the mountains, requires nothing more than following a line on a screen, without having recourse to the rest of the messy world around us.

This book has been written in homage to the pioneers who made it all possible, not least the unsung military surveyors – mountaineering's first true protagonists – who hauled heavy instrumentation to almost insurmountable summits in all kinds of weathers to take accurate measurements of the terrain and produce a map.

* * *

Endnotes

1 *Tsurugidake: ten no ki*, the Japanese title of the film, is based on the novel written by Nitta Jiro.

2 The surveyors battled first and foremost against the elements, the rivalry with the Japanese Alpine Club was woven into the storyline. Astonishingly, the discovery of the ancient relics is true.

Astronomy

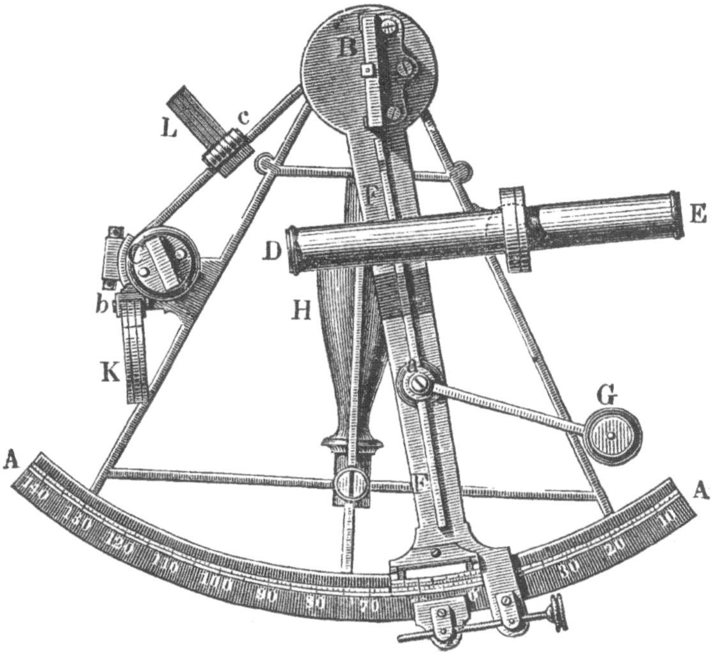

A marine sextant.

Early global positioning systems to pinpoint a fixed location on the Earth's surface relied on celestial observation. The study of the motions of the sun, the moon, the planets and the stars also led the Ancient Greeks to correctly believe that the Earth was a spherical object and they made early attempts to establish its exact dimensions.

Thus, the science of geodesy was born. Its aim was to determine the shape and size of the Earth, as well as to define the exact position of geographical locations. The founding father of geodesy is generally accepted to be Eratosthenes of Cyrene (c.276-c.195BCE), a Greek scholar who became the chief librarian at the Great Library in Alexandria. It is said that his great Eureka moment came when - in around 240 BCE - he discovered that on the day of the northern hemisphere's summer solstice, the rays from the midday sun shone directly into the bottom of a well in Syene, a city in southern Egypt. Eratosthenes observed this was different in Alexandria, about 800 kilometres due north of Syene, where the shadow cast by an upright pole at noon on the longest day of the year was seven degrees out of perpendicular, roughly one-fiftieth of a circle.

Armed with the figures he had at his disposal, the chief librarian was able to make a reasoned calculation of the size of the Earth at the modern equivalent of 39,375 kilometres. We now know he was out by 1.4 percent, but by the standards of the day - given the absence of accurate instrumentation - his calculation was pretty much spot on, the actual circumference of the Earth at the equator now being calculated at 40,075 kilometres. Eratosthenes' measurements also established that Alexandria was roughly seven degrees due north of Syene which, if the sun was overhead on June 21, lay on the Tropic of Cancer at 23.5 degrees north of the equator. It may be a moot point that Syene actually lies 24 degrees north of the equator (and Alexandria 31 degrees), but it confirmed a principle for fixing latitude by measuring the angle of the sun.

Although it may have intrigued the early geodesists, latitude and longitude positioning was not a consuming interest for land travellers. It was on the open sea, devoid of any fixed landmarks, that knowing your location could make the difference between life and death: celestial bodies were all early seafarers had to go by. Identifying latitude could be done by calibrating the height of the midday sun or the Polestar in relation to the horizon. Even in choppy seas, fog and cloud, fixing latitudinal position was not an insurmountable challenge. Determining longitudinal position was much more problematic as it was dependent on knowing the exact local time in the middle of an ocean *and* the time at fixed meridian, for example, at the ship's home port.[1] In the age of satellite navigation and electronic clocks, this might seem straightforward, but to ocean-going navigators without accurate timekeeping equipment it was hit and miss.

Galileo proposed using the moons of Jupiter to solve this dilemma. Using a telescope, he discovered that the planet's natural satellites orbited with such predictable regularity that they worked like a clock. From a moving ship, sighting Jupiter's moons through an eyepiece was always going to be a challenge however, and it wasn't until a marine chronometer – or reliably accurate clock – that could be used on board a ship was invented that the enigma of longitude at sea was finally solved.[2]

In the meantime, the scientific advances in astronomical observations which had been taking place since the time of Galileo allowed astronomers and surveyors to pinpoint exact locations on land. For example, by 1684 the French *Académie des Sciences* had mapped the coastal outline of France using this principle. The *Académie* was able to produce an outline map of the country which was more accurate than anything seen before. Later, fixed telescopes (including zenith sectors) were used in mountain terrain to pinpoint key geographical locations to corroborate data arrived at by triangulation.

A 16th-century Turkish illustration of a quadrant.

Instrumentation for measuring the angles of celestial objects

To calculate the exact altitude of the sun, the moon and the stars as they passed the observer's meridian, early astronomers used the quadrant. This was a quarter circle instrument - a kind of glorified classroom protractor - that was able to measure angles through 90 degrees, from the horizon to the zenith, the point vertically above the observer.

Quadrants - and adaptations which improved upon its basic design - were particularly valuable at sea for determining the latitude of the vessel. The Davis backstaff[3], for example, was a device for measuring the height of the sun by the projection of its shadow, but it was impractical - or rather, useless - for observing other celestial objects, which cast no shadow. The sextant, first implemented in the mid-18th century, overcame this problem and could also be used at nighttime for measuring the angle of the moon and stars, without also requiring a completely steady aim, which could otherwise prove difficult on a moving vessel.

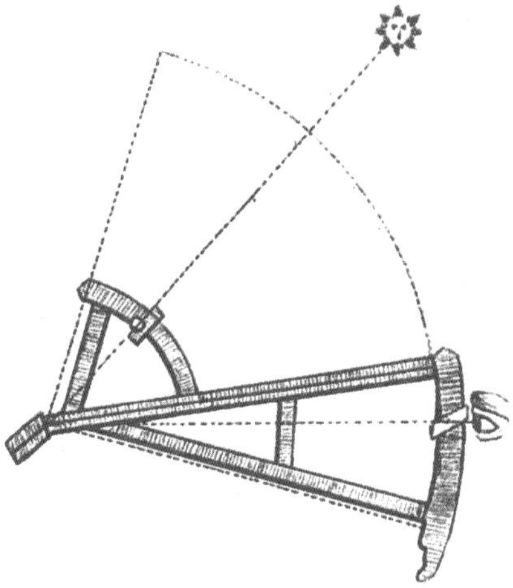

A 17th-century backstaff.

On land, the angle of a celestial object could be more easily determined by the use of fixed telescopes. Amongst other things, the invention of the zenith sector, or zenith telescope, in the late 17th century enabled a more exact calculation of one's latitudinal position. The telescope was aligned with the perpendicular using a plumb line to establish the zenith position. The telescope could then be rotated and sighted on a star and the angular distance measured to calibrate the star's exact position relative to the zenith.

One of its advantages over other instruments is that, being pointed upwards, it suffered very little from atmospheric refraction. One such mobile zenith telescope, for example, was used to establish exact latitudes of many locations in the Principal Triangulation of Great Britain in the 19th century.

Jesse Ramsden's zenith telescope was used in the Principal Triangulation of Great Britain which was carried out between 1791 and 1853.

Endnotes

1. Greenwich became the universally adopted meridian, or longitude zero, by an international convention signed in 1884.
2. The first accurate marine chronometers came into being in the mid-18th century
3. Named after John Davis, a leading English mariner during the late 16th century.

Triangulation

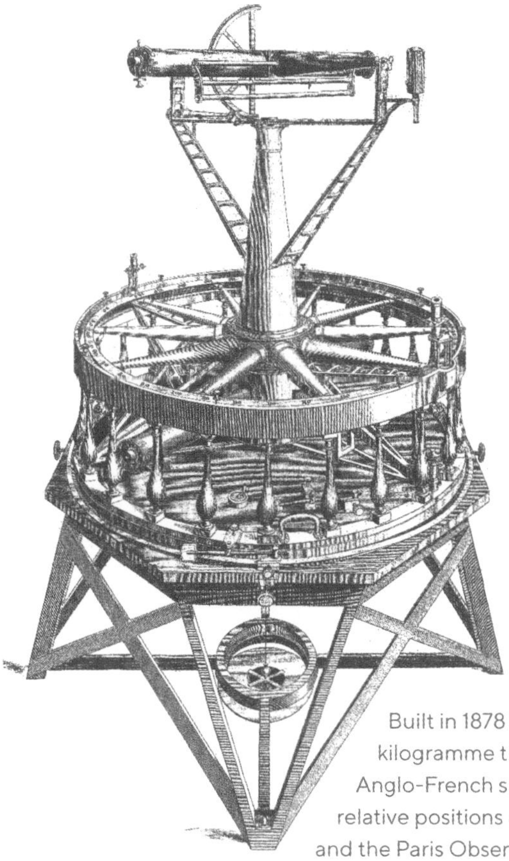

Built in 1878 by Jesse Ramsden, this 100-kilogramme theodolite was used for the Anglo-French survey of 1784-1790 to measure the relative positions of the Royal Greenwich Observatory and the Paris Observatory by means of triangulation.

Triangulation is one of the pillars on which modern cartography is founded. It is a simple and accurate method of surveying by which distances and angles can be measured over large expanses of country through the application of geometric principles. Whereas triangulation as a tool for surveying has been known about since ancient times, it was only after the rediscovery of ideas and achievements of classical antiquity that it came to be used systematically for mapping purposes.

Most of us who have gone through secondary school will have a grasp of the fundamentals of geometry, a branch of mathematics that is concerned with the properties and relationships of lines, angles, curves and shapes. Trigonometry, more specifically deals with the properties associated with triangles – for example, the relationships between the lengths and angles of a triangle – its name being derived from the Greek for 'triangle' (τρίγωνον - trígōnon) and 'to measure' (μέτρον - metron). Who doesn't remember the sine, the cosine and the tangent and functions of trigonometry as part of the maths curriculum? Whilst trigonometry looks after the theory, triangulation is the method by which its principles are applied in practice. Especially so in modern cartography.

Essentially, triangulation enables us to calculate the length of two unknown sides of a triangle if one of its sides and the two internal angles relative to the third point of the triangle are known. If the distance between two points – say a clock tower (A) at one end and radio mast (B) at the other – is known, then using a simple formula, the exact position of a visible point C (i.e. a prominent hilltop) can be precisely determined by making a triangle, as long as the internal angles at each end of the baseline (from A to B), relative to C are known. Once the exact distance of each of the three sides of the triangle has been calculated, a whole chain of new triangles can be created, each side of the triangle acting as a new baseline.

Triangulation was used in the ancient Greek, Arab and Chinese cultures to measure heights, slopes and angles and was even used for basic mapping. However, it wasn't until the late eighteenth century that triangulation came into its own on a large scale for cartographic purposes.

Early scientific application of triangulation began in the Low Countries. Dutch mathematician Gemma Frisius (1508-1555) first described the concept by measuring a rudimentary baseline between Brussels and Antwerp and calculating angles and distances to church towers in places like Ghent, Mechelen, Leuven and Bergen-op-Zoom. Later, in 1615, Dutch astronomer and mathematician, Willebrord Snellius (1580-1626), working at the University of Leiden in the Netherlands, measured one degree of latitude between the towns of Alkmaar and Breda. He formed a network of triangles linked almost exclusively by church spires, the only tall buildings at that time in the west of the country. Snellius calculated the distance between the Sint-Laurenskerk in Alkmaar and the Grote Kerk in Breda at 112.05 kilometres, which was surprisingly accurate given that the actual distance is 116.1 kilometres.

The Cassinis in France were the first to apply triangulation on a wide scale, first by delineating a meridian line north and south of Paris and later by setting up a nationwide network of 800 triangles with additional baselines to create the first accurate map of France.

Triangulation soon became the established norm for mapping and was widely adopted by other European

If the length (A to B) of one side of a triangle (the baseline) is known, and the respective angles from A and B to point C are accurately measured, it is possible to calculate the lengths of the two other sides of the triangle (A to C and B to C).

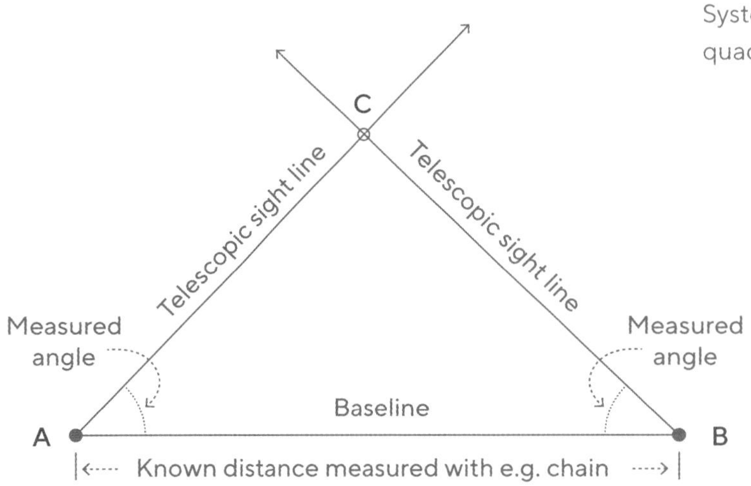

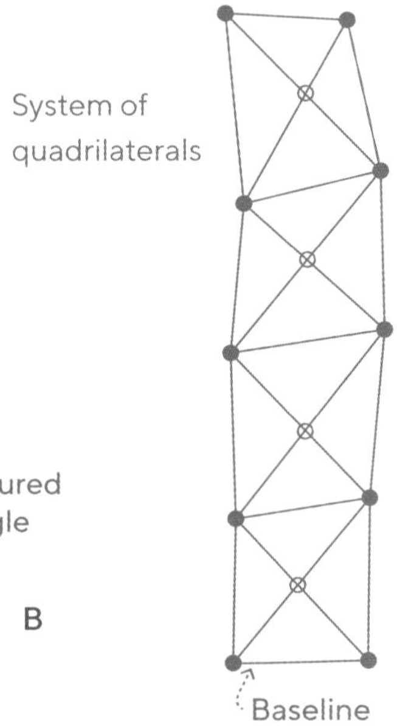

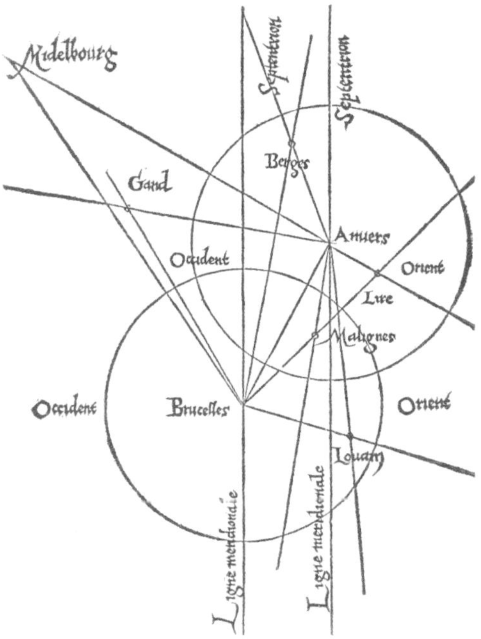

Gemma Frisius' 1533 diagram introducing the concept of triangulation.

countries. Instruments for baseline measurements (rods and chains) and angle calibrations in the field (theodolites) became ever more sophisticated, producing increasingly accurate networks of triangles.

Some nationwide mapping surveys ultimately ended up in mountainous regions, such as the Alps and the Pyrenees in France. Here, to get the best visible sights, triangulation stations were generally positioned on peaks and prominent hilltops. Often, it meant that heavy equipment had to be hauled across difficult terrain by teams of surveyors and their assistants. At other times, raised platforms would have to be constructed to provide a line of sight not obscured by trees or buildings.

One of the greatest ever undertakings in triangulation was the Great Trigonometrical Survey of India, commenced in 1802 by George Lambton and continued under his successor George Everest in 1823, culminating in the mapping of the Himalayas (see chapter 11).

In Great Britain, triangulation points commonly known as trig points have become icons of the British landscape. Around 6,500 of these concrete pillars were positioned strategically across the country as part of a project of retriangulation between 1935 and 1962. As trig points are located on some of Britain's highest hills, 'trig bagging' has become a popular pursuit amongst hill walkers in recent decades.

Instrumentation for measuring angles in triangulation

Angular measurements were absolutely crucial to triangulation surveys. Instruments that incorporated telescopes to measure angles and heights became more common in the 18th century. The general term used for describing such instruments that measure angles between designated visible points, whether horizontal of vertical, is the theodolite. Essentially, this is a telescope which is mounted on a base or tripod, rotated along its horizontal and vertical axes and then fixed on two sights to provide a readout of the angle between the two. Among the early theodolites were those manufactured by Jesse Ramsden[1] for use in the first triangulations of Britain and India, which might weigh anything up to 100 kilograms. Slowly but surely, precision engineering was able to minimise their size significantly. At the time Ramsden was building these huge theodolites, the French were using their own favoured instrument for triangulation purposes, the repeating circle. This instrument consisted of two telescopes likewise mounted on a fixed base with a shared axis. The telescopes could respectively be fixed on two distant points and the angle measured between the two and this exercise could be repeated multiple times to increase the accuracy of the measurements, hence the 'repeating circle'.

A plane table is an instrument by which a map can be sketched in the field without the need for other angle-measuring equipment (see illustration on page 117). The plane table is basically a flat drawing board with a sheet of paper attached, mounted horizontally on a tripod. An alidade, or ruler with sights, is used to plot the bearings of visible landmarks. Distances to these landmarks are measured out manually and the information is then transferred directly onto the paper to produce what looks like spokes of different lengths radiating from the hub of a wheel. A plane table is best suited to small-scale surveying and often used to fill in details between fixed triangulation stations, without the need for unwieldy surveying instruments.

Measuring distance

The perambulator (also known as a surveyor's wheel or odometer) was a wheeled device used to record distance travelled. They were used primarily to measure road distances (or itineraries) and often used in conjunction with a compass to measure changes in the course of a road. They were not designed for calculating precise measurements of distance: for example, the accuracy of the wheel's motion might be affected by changes in the road surface, such as mud or sand.

For more accurate measurement of distance, especially in the calibration of baselines for triangulation, chains or rods would be used. Metal 'chains', connected by 'links', were of a standard length, the exact unit of measurement for which differed per region.[2] It is said that linked chains were first introduced in the Netherlands for land surveying purposes in the 16th century. Fixed baselines, designating the all-important first side of a triangle in a triangulation survey, required more exact calibration. Chains were susceptible to stretching, and such was the concern that early geodesists used wooden rods, but these too were highly sensitive to changes in temperature and humidity which affected their length. William Roy, head of the Ordnance Survey at the time, experimented with glass rods for the Greenwich-Paris triangulation in 1785, before finally settling on a high-precision metal chain.

Perambulators and/or chains in combination with compasses and/or theodolites would be used in 'traverse surveying' which combined linear and angular measurements. Plane tabling could also add to the accuracy in the mapping of the surrounding topography. This type of surveying was in widespread use in the 18th century before the systematic use of triangulation.

Measuring height

The theodolite could also be used to measure angles in elevation.[3] In the field however, other methods could be used to calculate height without the aid of such high-precision instrumentation.

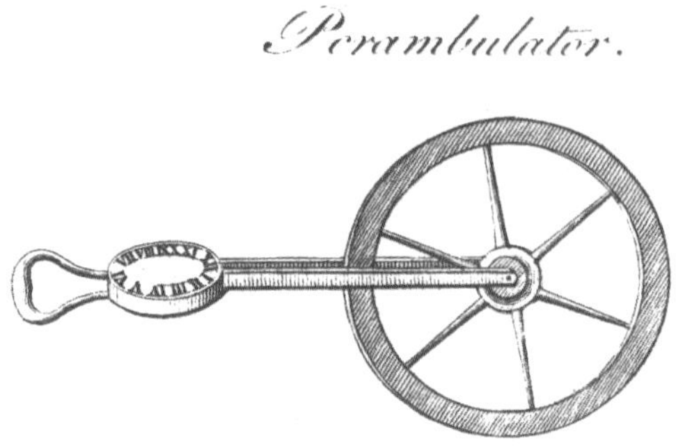

A perambulator or surveyor's wheel.

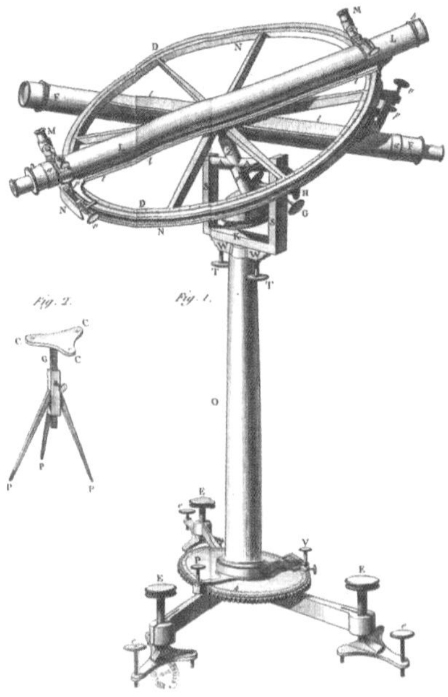

The French 'theodolite' of choice was the repeating circle. It was used, for example, to measure the meridian arc from Dunkirk to Barcelona by Jean Baptiste Delambre and Pierre Méchain (see chapter 10: Napoleon's Triangles).

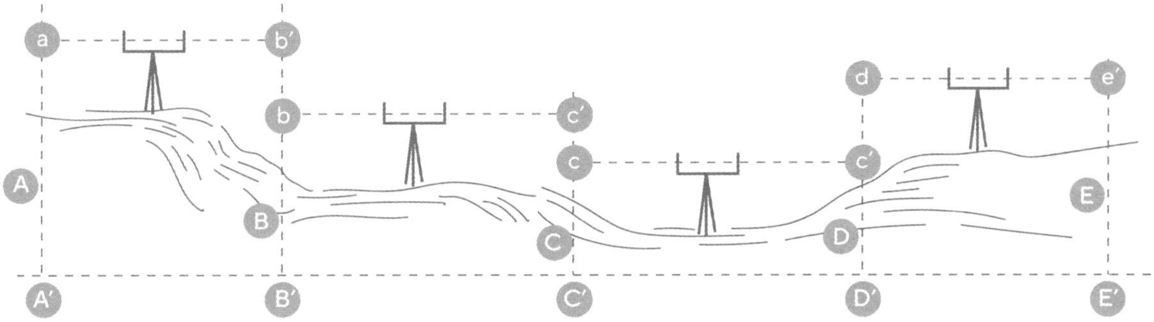

An example of levelling.

Altitude can also be determined by measuring atmospheric pressure. As altitude increases, atmospheric pressure decreases, so this means that the density of air at high altitudes (i.e. in mountainous environments) is thinner than air at sea level. The instrument for measuring changes in air pressure is the barometer, and the standard unit of air pressure is the millibar.[4] Whereas the barometer will give readings in millibars, an altimeter will convert this to altitude (taking into account that air pressure does not decrease in a linear fashion).

Atmospheric pressure also influences the boiling point of water: as atmospheric pressure decreases, so does the temperature at which water boils. Water will boil when its vapour pressure is equal to that of atmospheric pressure. At sea level, water boils at 100 degrees Celsius. Roughly speaking, for every 300 metres elevation, water will boil at 1 degree Celsius less: on the top of Mont Blanc (4,808 metres) water will boil at 83.5 degrees Celsius and at the summit of Mount Everest (at 8,848 metres) at 68 degrees Celsius.

On uneven ground, working in the field, levelling was the common method for measuring relative heights, for example, along the course of a road, but on other terrain too. When two points of differing height need to be compared, vertical staffs or poles with graduated markings are erected at both of these points. Using a horizontal line of sight (calibrated by means of a spirit level) between the two poles, the difference in height can be calculated. This process could be repeated at multiple points, using a reference height (e.g. the first pole) as the so-called 'datum' (whose height above sea level may or may not be known).

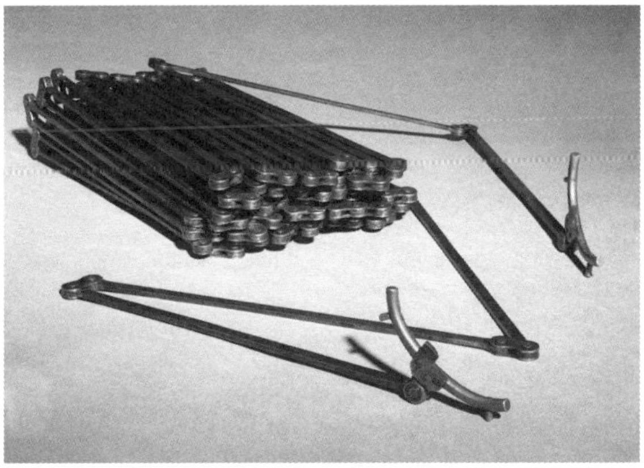

Jesse Ramsden's chain was designed to measure distances, such as baselines, accurately. The chain consisted of 100 links of one foot each (1 foot = 30.48 centimetres).

Endnotes

1 Jesse Ramsden (1735-1800) was an inventor of instruments for surveying, including theodolites, chains and telescopes.
2 In Britain, the 'chain' – measuring 22 yards with 100 links – became a standard unit of length, being one-tenth of a furlong.
3 This is how the high Himalayan peaks in Nepal were measured from triangulation stations in India as part of the Great Trigonometrical Survey.
4 1 millibar is equal to a pressure of 1.0197162129779 grams per square centimetre and at sea level the average pressure is 1013.25 millibars.

Photogrammetry

This photograph, showing Mount Assiniboine (3,618 m) was taken in 1913 by A.O. (Arthur) Wheeler as part of the Inter-Provincial Boundary Survey between Alberta and British Columbia in Canada. The survey was one of many Canadian surveying projects to implement the technique of photogrammetry, often in mountain terrain. Today, over a hundred years later, it is possible to use the thousands of photos that have been archived, such as this one, for scientific research purposes.

By the end of the nineteenth century, advances in technology meant that conventional trigonometrical surveys could be more readily carried out by means of photogrammetric triangulation. Also known as phototopography, this innovation lent itself explicitly to mountain surveying: instead of using theodolites, surveyors could place a camera on a tripod at fixed survey stations, direct the lens at various points along the horizon and take panoramic shots of the surrounding peaks and topographical features. At the end of the season they would pack up their gear, return with their photographic data to the warmth of their offices and convert everything into maps. Compared to conventional surveying, larger areas could now be mapped more accurately, with added savings in time and money.

Practical photogrammetry on the ground - so-called terrestrial photogrammetry - was first applied for mapping in the 1880s. It was a technique that was widely adopted in the uncharted territories of the Canadian Rockies as the Canadian Pacific Railroad pushed its way westwards, opening up this vast region for tourism and the exploitation of natural resources. On the back of the expertise acquired, Canadian surveyors were much in demand elsewhere, such as Oliver Wheeler, who was responsible for creating the first detailed map of Everest using photogrammetric techniques on the 1921 British reconnaissance mission (see chapter 15).

The German and Austrian Alpine Clubs - working in collaboration - were early converts to photogrammetry too, because of its practicability in difficult mountain terrain and they are still producing excellent maps today. Richard Finsterwalder (1899-1963) became a standard-bearer for the method, and his name is indelibly associated with maps produced through this collaboration, none more so than the 1:50,000 map

Swisstopo aerial photo of Matterhorn, 1956. The summit is on the right-hand edge of the photo, the shadow it casts clearly visible in the top right.

Richard Finsterwalder surveying using a camera, Nanga Parbat expedition, 1934.

of Nanga Parbat, published after the ill-fated 1934 German expedition to the mountain: it is still considered a masterpiece today.[1]

With the advent of aviation, from around 1920 onwards, the importance of terrestrial photogrammetry was gradually supplanted by aerial photogrammetry. Starting in 1926, the Swiss national mapping agency began taking aerial photographs of the country every year and by the late 1940s, aerial images had completely replaced terrestrial photographs. Self-evidently, aerial photography (as for photogrammetry in general) is dependent on clear and cloud-free weather conditions.

Vertical photos, however, do not replicate what is seen on the ground exactly: there is always some distortion towards the edge of the photograph, but also in respect of mountain tops, which, because of their greater proximity, appear to be at a larger scale than the valleys, which are further away from the camera. In order to get a more accurate three-dimensional perspective, a series of overlapping photos are taken as the aircraft travels along a straight flight path overhead. The stereoscopic effect of combining these photos provides the cartographer with a more accurate depiction of the topography, especially in mountain terrain. The most iconic mountain map produced using this method is Bradford Washburn's 1988 map of Mount Everest (see page 204), which combined photographs with ground-based data taken from historical surveys. It was created in collaboration with the Swiss mapping agency and the National Geographic Society.

Nowadays, aerial photography can also include satellite imagery and, generally, any mapping that is carried out from above the surface of the Earth these days is referred to as 'remote sensing'. Typically, satellites used for mapping orbit the Earth at altitudes of 300 to 1,400 kilometres above the Earth and they can measure a lot more than the topography alone. Clearly, the mapping of the Earth's surface has come a long way since Ptolemy!

Endnotes

1 In 1948, Finsterwalder became a professor of photogrammetry at Munich's Technical University. Photogrammetric surveying was used also by the DAV/ÖAV to produce detailed maps of the South American Andes and the Himalayas, updated and still in use today.

Depiction of relief

This 1932 map "Jungfraugruppe und Aletschgletscher" designed by Eduard Imhof, is illuminated from the south.

Depiction of relief, especially where mountainous terrain has to be reduced from three dimensions onto a two-dimensional, plane surface is one of cartography's abiding problems. In medieval times, mountains were seen as largely irrelevant to the function of a map, providing obstacles to communication without having any economic or cultural significance and they were mapped as rows of molehills or cones along the edges. As interest in mountains developed over the centuries, sophisticated methods were developed to express elevation and slope angles on maps, the most important of which are discussed below.

Spot heights
At various points on a map, heights are marked which represent elevations relative to sea level on the Earth's surface. Whilst they provide definite and precise information, they give little indication about the general pattern of relief and have to be used in combination with other methods of relief depiction to be of any great value. They should be carefully selected where needed by the map user, for example on peaks and saddles, and at important locations, such as mountain huts or the terminal stations of cable cars, where they can provide valuable information for hill-climbers.

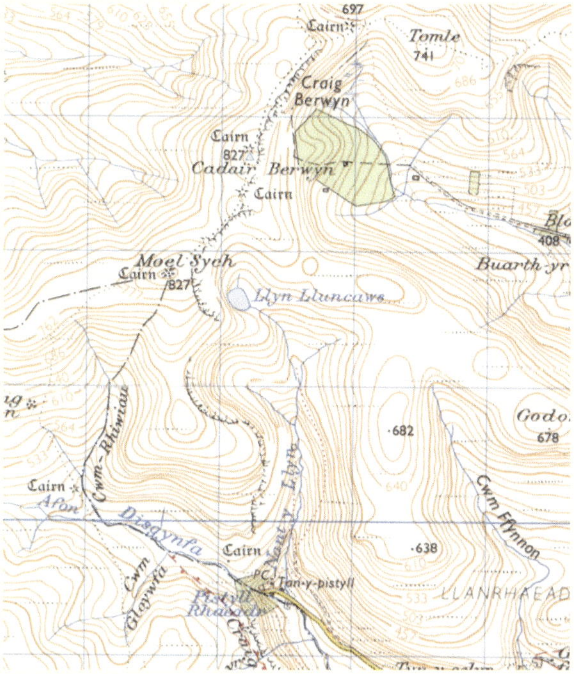

Spot heights on an 1:50,000 Ordnance Survey map of Wales, marking key points in the topography, such as summits and spurs, rock outcrops, but also road heights.

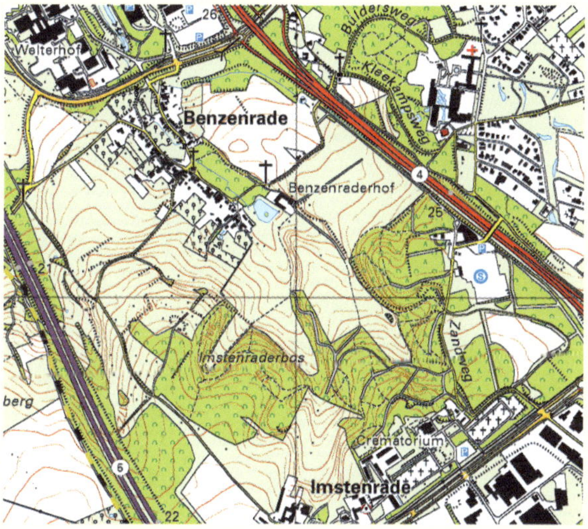

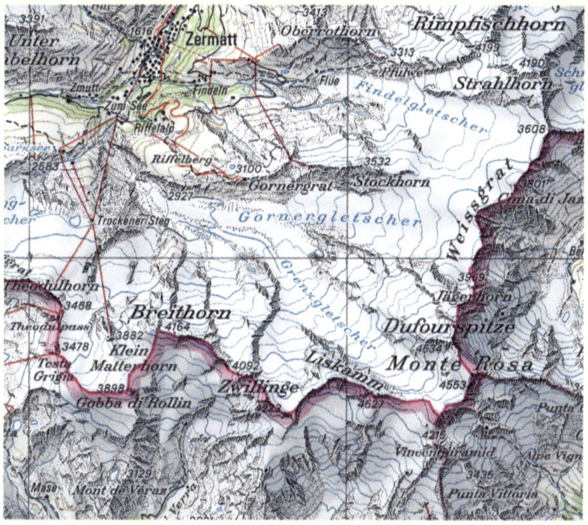

On the Dutch 1:25,000 topographic map (top), contours are spaced at 2.5-metre intervals, whereas on the swisstopo 1:100,000 map of the Monte Rosa region in Switzerland, contour intervals are every 100 metres in height.

Contour lines

Contour lines link all points on the Earth's surface that are at the same elevation above sea level. In the past, contour lines were interpolated by joining spot heights or through the skilled eye of a surveyor. Nowadays photogrammetry with aerial photographs and stereoplotting machines provide extremely accurate contouring.

The choice of contour interval, that is, the vertical distance between the two contour lines on a map, is critical. It would seem logical to assume that the smaller

Example of detailed rock hachuring on a Swiss topographic map.

the interval, the more accurate the map, but this does not necessarily improve legibility. So, the smaller the contour interval, the more crowded the information on a map becomes: contour intervals of five metres might be fine in the Netherlands (even in its hillier regions like Zuid-Limburg), but they would not be useful in a mountainous country like Switzerland.

This is equally true of scale: the smaller the map's scale, the more condensed contour lines become. Some mainly smaller-scale maps combined contours with contour-layer colouring, where different colours between different contour lines indicate the relative heights and depths in a landscape: green for lowland regions, increasingly darker shades of brown for moors and higher-lying ground, merging into purple and then white for mountain tops.

Contour lines can be used in different colours to represent different terrain. On maps produced by swisstopo, the Swiss national mapping agency, black is used for bare rock and scree, blue for ice and underwater contours, and brown for earth-covered ground.

Hachures

Hachures are lines drawn down a slope to indicate gradient: the steeper the slope, the thicker and shorter the hachure. Hachuring was developed in the early 19th century and was commonly used to depict relief in maps until the mid-twentieth century, when contouring became more common. This could be useful even in lowland mapping to indicate undulations in the topography. In more mountainous regions, shadow hachuring could be used to produce a better three-dimensional effect with more finely drawn lines on the 'illuminated' slopes of mountains – generally the north-west sides – and more heavily on the slopes in 'shadow'. The most significant example of this type of hachuring is the Dufour map of Switzerland. Hachuring also lent itself to the copper-engraving techniques that were used for printing maps. Hachuring was gradually superseded when contouring became more accurate and popular in the 20th century.

Relief shading

Relief shading introduced a more comprehensive method of shadowing mentioned above. Important to this technique was a choice of light direction, with most contemporary maps choosing to have a light source in the upper-left of the map. Essentially, it is the mapmaker's *trompe-l'œil*, producing an optical illusion which gives a three-dimensional appearance to a two-dimensional surface, highly applicable to mountainous and upland regions. Entire theses have been given over to the theory of relief shading and perhaps the most notable figure in this respect is Eduard Imhof (see pages 191-193).

Rock depiction

The depiction of extreme terrain could not be solved solely through hachuring or contours, so methods were developed to illustrate steep cliffs and rock faces. A traditional form of rock portrayal is shaded rock hachuring. Until recently, this had been done manually and is a highly skilled craft requiring intensive field work. Good aerial and land-based photographs taken at different times of the day are indispensable for this.

Example of shadow hachuring used to depict the Matterhorn on the Dufour 1:100,000 map of Switzerland. The Dufour Map was published between 1845 and 1865

Mapping the Mountains

An 1866 print by Heinrich Leutemann of Hannibal Crossing the Alps.

Before the maps

Hostile mountain environments did not easily lend themselves to economic development and mapping

Throughout most of human history, mountain regions were places to avoid. In the beginning, their sheer majesty and awe must undoubtedly have captured the imaginations of those living amongst their midst or passing through: their inhospitable conditions and barren wastes offered precious little in terms of economic sustenance. Mountains formed an insuperable barrier to trade and development for those living on either side, but conversely this impregnability could also provide a sanctuary from hostile aggressors. All things considered, there was very little reason for those living in primitive societies to venture into this perilous, unknown void.

Yet their ramparts were never completely impenetrable. In the crumpled labyrinth of peaks and valleys formed by millions of years of uplift and denudation, adventurers would cautiously navigate their way across the jagged watersheds, seeking out the path of least resistance. Surmountable mountain passes provided openings for enterprising traders, as well as belligerent warlords or fleeing armies. For centuries, those who wished to navigate their way across the mountains would hire local inhabitants to lead them safely through the perils they faced. Over time, the routes that presented the fewest obstacles would find favour with travellers, and an infrastructure would develop around them.

Archaeological evidence has established that the Alps, for example, had been settled long before written records began. In 1991, the mummified remains of a man who had lived around 3350 to 3100 BCE were discovered in a body of ice high in the Ötztal Alps, just south of the present-day Austrian-Italian border. Archaeologists dubbed this iceman Ötzi, and findings gleaned over the last thirty or so years point to the fact that he was a high-altitude herdsman who was murdered by other hostile mountain dwellers.

The various Celtic tribes that later occupied the Alpine valleys always represented a thorn in the side of the European superpowers of the day. Polybius, the Greek historian, reported that Hannibal's attempts to traverse the Alps in 218 BCE were initially thwarted by a fierce mountain tribe known as the Allobroges. Ultimately, the Carthaginian general would cross the divide with his elephants, doubtless guided by more friendly mountain dwellers, and set in motion the Second Punic War. And before they were properly able to consolidate their realms north of the Alps, the Romans first had to subdue the troublesome tribes that occupied the regions close to the traversable mountain passes.

Mountain passes were the pivots around which movement across the Alps revolved. When it came from getting from A to B, the peaks themselves were of little interest, only the openings that led from one valley to the next were relevant to travellers. When the Romans eventually broke the resistance of the Alpine tribes they opened up military and trade routes across the Alps. With the possible exception of the Great Saint Bernard, the passes they pioneered were not necessarily those we consider the most important today.

Although the Romans revolutionised communications across the Alps through their superior engineering, they lacked the technology to overcome the more immovable obstacles that lay in their path, for example, on the approaches to some of the lower mountain passes.[1] The Peutinger map, a medieval reproduction of an earlier Roman route map shows three main routes across the central Alpine chain: the Grand Saint Bernard; the Splügen; and the Reschen.[2] Other key passes across the Alpine watershed were at Montgènevre and the Little Saint Bernard (on the present-day French-Italian border) and at Tarvisio in the Eastern Alps (near the Austrian-Italian border).

Identifying where Hannibal and his war elephants crossed the Alps in 218 BCE remains an unsolved mystery. One possible crossing point is thought to be Col de la Traversette (2,947 m) in the Cottian Alps.

Endnotes

1 For example, the Gondo gorge near Domodossola on the southern approaches to the Simplon and the Eisack gorge north of Bolzano/Bozen in the direction of the Brenner.

2 In fact modern-day commentators believe that the Septimer Pass (rather than the Splügen) was more widely used by the Romans for the same reasons as indicated above. The Septimer Pass is 30 kilometres to the south-east of the Splügen.

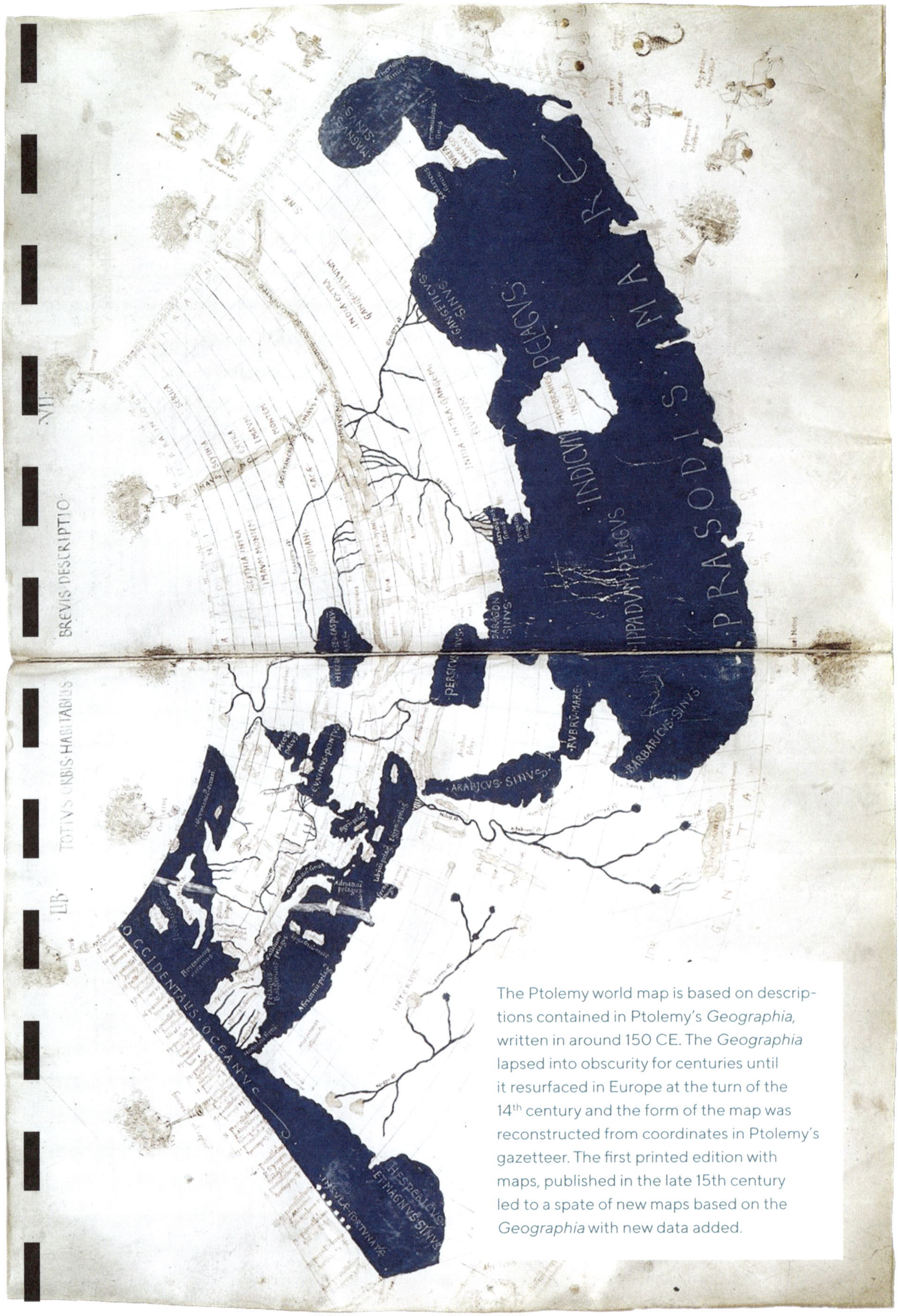

The Ptolemy world map is based on descriptions contained in Ptolemy's *Geographia,* written in around 150 CE. The *Geographia* lapsed into obscurity for centuries until it resurfaced in Europe at the turn of the 14th century and the form of the map was reconstructed from coordinates in Ptolemy's gazetteer. The first printed edition with maps, published in the late 15th century led to a spate of new maps based on the *Geographia* with new data added.

Geography comes of age

One of cartography's milestone moments took place in the second century CE in Alexandria

The city of Alexandria, which lies on the Mediterranean shoreline near to where the Nile empties into the sea, was founded by Alexander the Great in 331 BCE. Apart from its soaring lighthouse – one of the Seven Wonders of the World – Alexandria was known primarily for its Great Library, the most enlightened centre of learning in the ancient world. The library functioned as a repository, collecting, transcribing and translating texts from all over the known world. Its accumulation of information was paralleled only by the number of influential scholars that passed through its illustrious doors and, as far as geography was concerned, none was quite so influential as Ptolemy.

Claudius Ptolemy, who lived from around 90 to 168 CE, is generally acknowledged to be the father of geography.[1] He was preceded by several other accomplished scholars in its study at the Library in Alexandria. The most notable of these was Eratosthenes (c.276-c.195 BCE), who was the first known geodesist to make an estimate of the size of the Earth's circumference based on scientific reasoning (and with an amazing degree of precision by today's standards). It was Ptolemy however, who developed on the ideas of his predecessors and who has been credited with producing the earliest cartographic 'tool-kit'.

Ptolemy was possibly the first to describe a scientific method for accurate mapping. The underlying principle was that reliable maps could only be produced on the basis of astronomical observations. To do this, Ptolemy amassed a dataset (or gazetteer) of more than 8,000 coordinates of towns and natural features in the known world (Europe, Asia and Africa) and had them plotted on a grid based on their latitude and longitude readings.

His grid was not rectangular, but took into account the curvature of the earth: whereas lines of latitude ran equidistant to each other, the distance between lines of longitude tapered as they approached the Poles. In this sense, Ptolemy was one of the first to advance the concept of map projection. Latitudes were much easier to define, for example, by measuring the height or angle at midday, than longitudes. The latter depended on knowing their exact time in relation to a prime meridian (i.e. the 0° line of longitude), based on the Earth as an almost perfect sphere rotating through 360° in 24 hours.

In the Ancient World, no accurate timekeeping devices existed for doing this and the only known method was in the timing of lunar eclipses, which are notoriously few and far between. The measurement of longitude would remain a major headache for navigators until Galileo discovered the moons of Jupiter in the early 17th century.[2]

Having already compiled the *Almagest*, a major work on astronomy, Ptolemy went on to compose the *Geographia*, a three-part treatise on geography. The first part consisted of an introduction on how to create maps based on his hypotheses; the second comprised the list of the 8,000 locations throughout the Roman world and beyond with their geographical coordinates; part three was an atlas showing a map of the known world, plus 26 regional ones.

Though the science in part one of his treatise has stood the test of time, Ptolemy's practical implementation of his principles has not. Defining distances between lines of longitude relied in the first place on an accurate measurement of the circumference of the earth: Ptolemy estimated it to be three-quarters of its actual size. This underestimation had the effects of downsizing these distances. Secondly, creating a map of the known world also required a reliable calculation of coordinates. For his 8,000-plus locations, he used data collected by another geographer, Marinus of Tyre[3] (c.70-120 CE) and supplemented these by travellers' accounts, which more often than not were unreliable.

The *Geographia* was written around 150 CE, at a time when the Great Library was already diminishing in status. Alexandria could be a volatile place during the Roman period; the city was ravaged and the library and its contents ransacked on several occasions. It is believed that by the end of the fourth century CE the library had been razed to the ground, once and for all.

It would be another thousand years before the Western world took any cognisance of Ptolemy's *Geographia*. It resurfaced in the Middle Ages, after a text was rediscovered by Maximus Planudes around 1300. The Greek text was eventually translated into Latin, but it was not until 1477 that a printed edition was published with maps, based on the data in Ptolemy's original gazetteer. Many more versions and translations of the *Geographia* appeared in the 16th century, their maps gradually improving, with more and more information being added to the existing dataset, and eventually expanding to include newly explored areas of the world.

The maps produced on the basis of Ptolemy's *Geographia* were a significant advancement on existing ones that were available in Europe during the Middle Ages. Not that medieval maps were widely available anyway before the advent of printing, which happened to coincide with this newly found interest in the *Geographia*. The *tabulae antiquae*, based on Ptolemy's data, bear an uncanny resemblance to the maps we are familiar with today. Even so, there are still many inaccuracies, primarily ascribable to Ptolemy's errors in measurement. For example, the island of Britain is ludicrously misshapen, with Scotland, instead of being 'upright', bending itself over backwards in relation to England and Wales. It is also a Eurocentric map (a fact of which Ptolemy was fully aware): the further the distances from the Mediterranean, the more erroneous the outlines become. However, as more and more *tabulae novae* started to appear and cartography improved, these errors were gradually ironed out.

Ptolemy.

Mapmakers of the early Renaissance who applied Ptolemaic principles were primarily interested in world and continental maps, rather than zooming in on smaller (mountain) regions. Sebastian Münster, on the other hand, was a German-Swiss mapmaker who was greatly influenced by Ptolemy but who also created several regional maps. Working in the 1540s, he included two new *tabulae novae* of Valais on his maps, one devoted to the mountains north of the Swiss canton, and another to the Pennine Alps on the southern border with Aosta. His maps of Valais and his 1540 map of Switzerland show important topographical features, such as lakes, rivers and mountains. Likewise, his regional *tabula nova* of Italy is richly embellished with depictions of the Alps and the Apennines running down the spine of the peninsula.

As for Ptolemy, his impact on cartography should not be underrated. Up until the late Middle Ages, the Western world had not drawn on any plausible principles for mapping. It is somewhat ironic that it took the rediscovery of Ptolemy, active a thousand years earlier, to kick-start the modern era of cartography, one which was based on science and would serve the needs of exploration, colonisation and economic and social advancement in the centuries that followed.

Sebastian Münster's 1552 map of Italy is richly embellished with depictions of the Alps in the north and the Apennines running down the spine of the peninsula.

Endnotes

1 His origins are unknown. The Claudius in his name implies that he may have been a Roman citizen, as Alexandria was part of the Roman Empire at the time. But he also wrote in Ancient Greek and drew on Babylonian knowledge.

2 See section on Astronomy

3 Marinus of Tyre was a geography scholar whose ideas and practices had a considerable influence on Ptolemy. He is credited with having introduced the equirectangular projection. He was also the first to have referred to the southern polar region as Antarctica (i.e. the opposite of the Arctic).

XIX· NOVA TABVLA·
Carinthia
Croatia
Sagabria
SERVIA
BOSSINA
Turcia
SCLAVONIA
DALMATIA
Campus Cassonius
Venetiæ
Aquileia
Friaul
Histria
Pola
Mare Adriaticum
Adria
Ferraria
Padua
Imola
Ravena
Cremona
Pisa
Elba
Ancona
Scardona
Spalato
Melada
Aquila
Apulia
Roma
Caieta
Aleria
Mare Tyrrhenum
Ponce
Neapol
Salerno
Terra
Laboris
Taranto
Otronto
Hydrunti
Albania
Corfun
MARE MEDITERRANEVM
Lipari
Calabria
Aetna
SICILIA
Trapani
Gergento
Syracusa

On the itinerary

On the itinerary

A minimalist map showing the vast network of roads the Romans built includes those they used to cross the great Alpine passes

Monaco has been the scene of many notable victories – some at the roulette table, others on the race track – but none was quite so comprehensive as that which is commemorated at Tropaeum Alpium at the village of La Turbie, on the slopes overlooking the principality. This monument[1] - 'Trophy of the Alps' - was built in 7-6 BCE to honour the Emperor Augustus whose Roman armies cleared the Alps of all opposition, vanquishing innumerable Alpine tribes in their wake. At the time of its construction, the now semi-ruined monument stood 50 metres high and consisted of a tower surrounded by 24 columns and crowned by a statue of the emperor.[2] The triumphal arch lay at the western terminus of the *Via Julia Augustus* which originated in *Placentia* (modern Piacenza) and was later extended as far as *Arelate* (Arles). Before this conquest, this maritime route represented the only safe and secure way for the Romans to move their goods and troops to their northern provinces.

The Tropaeum Alpium overlooking the village of La Turbie, near Monaco.

Half a century before the monument's inauguration, Julius Caesar had already conquered vast swathes of Gaul, a region encompassing roughly the equivalent of present-day France and Belgium. However, the Alps, which remained in the hands of local tribes, formed a major logistical barrier. Gaining control of the mountain passes was a strategic necessity for the supply and thus expansion of Rome's Gallic possessions.

One of the key passes affording a more direct route north was the one we know today as the Great Saint Bernard.[3] At 2,469 metres above sea level, the pass is considerably higher than other notable passes in the Western Alps, such as Montgènevre (1,854 m), Mont Cenis (2,084 m), the Simplon (2,009 m) and Saint Gotthard (2,106 m). The advantage of the Great Saint Bernard however, lay in its more accessible approaches and the north-south orientation of the route linking the Aosta and Rhone valleys.

At the time of Julius Caesar, the Rhone valley to the north of the pass was commanded by the Veragri tribe, and on its southern flanks, in Val d'Aosta, the Salassi held sway. Merchants travelling over the pass, Roman or otherwise, would have been subject to tolls levied by these local tribes, and on occasions they would have even been attacked and robbed. Realising the importance of controlling such a strategic link to Rome's northern provinces, Caesar sent an expeditionary force from Gaul to take the pass in 57 BCE. Whilst briefly managing to establish a fortified camp at *Octodurus* (present-day Martigny), 40 kilometres north of the pass, gaining a permanent foothold proved impracticable. In the face of fierce local opposition and inadequate supply lines, the Romans beat a retreat.

It was left to Augustus, Caesar's adopted successor, to seize control of the whole Alpine region and ensure free access to the Roman territories north of the mountain chain. Heavy tolls exacted from passing troops and travellers by bellicose tribes would no longer be tolerated. The Salassi were eventually brought to their knees in 25 BCE and the Aosta valley which they had occupied was depopulated and militarised. Around 44,000 people were enslaved and a new colony was established at *August Praetoria Salassorum* (present-day Aosta).

The Great Saint Bernard had been secured once and for all. In an extraordinary demonstration of military might, in the space of just 20 years, the Romans had succeeded in pushing their northern frontiers beyond the Alpine watershed. By 5 BCE, Rome felt confident enough to claim total domination of the Alpine regions and the erection of their imposing trophy overlooking the Mediterranean at Monaco was confirmation of this.

It was not until the reign of Emperor Claudius (41-54 CE) that a road was completed across the Great Saint Bernard, on whose summit a *mansio* (an official lodging) and a temple, dedicated to *Jupiter Poeninus*, were constructed.[4] Roman engineers were ahead of their time when it came to building roads, bridges, aqueducts and the like, but roads at this height presented a completely different challenge. Steeper gradients, landslides and flooding, for example, never mind the fact that the pass would have been closed for six or seven months of the year because of snow.[5] After all, the Romans were notorious for building direct routes, rather than those that followed the natural contours of the terrain and, of course, tracks capable of taking wheeled vehicles were limited in their gradient. It is unclear whether

The Great Saint Bernard pass on the Swiss-Italian border, looking west.

a track able to carry such vehicles went all the way to the top and it was likely that some goods would have had to be transferred to mules to negotiate the steeper slopes. However, sections of a Roman road on the relatively flat summit of the pass have survived.

The settlement at *Augusta Praetoria* (Aosta) was of such strategic importance that it became a major garrison town whose sole *raison d'être* was to maintain control of the two key routes out of the valley and over the Alps (the Little[6] and the Great Saint Bernard passes respectively). Many ancient buildings from this period survive as monuments in the city today, as do sections of the ancient road to *Eporedia* (modern Ivrea), including a 220-metre length of road cut through bare rock at Donas (complete with Roman archway and milestone), a 35-metre single-span bridge at Point-Saint-Martin, and a six-kilometre aqueduct at Pont d'Äel, which supplied water to the garrison. Meanwhile, *Forum Claudii Vallensium* (modern Martigny) was of a similar strategic importance to Aosta on the northern side of the Great Saint Bernard.

It was not only in the Western Alps that Emperor Augustus was forging new corridors across the mountains. Archaeological findings point to a Roman military encampment on the Septimer Pass (2,310m) in present-day Graubünden in Switzerland. The camp dates from between 15 BCE and 16 CE and was used to control the pass and allow troops to move freely between Italy and the northern

The remains of the Roman Theatre at Aosta, with Monte Emilius (3,559m) in the background.

The Septimer Pass is one of twelve major Alpine passes, one of only two or three passes in the Graubünden Alps known to the Romans, and the most important one in this region until the middle of the first century. Further east are the Reschen and Brenner Passes which were likewise opened up by Augustus and his successors. The former formed part of the *Via Claudia Augusta*, a route likewise developed to improve troop movements to Raetia, which was later superseded in importance by the *Via Raetia* over the Brenner Pass in the 2nd century CE.

If the maps the Romans left behind are to be believed – and there are precious few – the mountain passes were key transport routes across the mountains. Nevertheless, for all their military might, expansionist drive, engineering prowess and attention to detail, the Romans left behind little detailed topographic information at all.[7] If it hadn't been for a redrawn copy of a map that surfaced in the Middle Ages, we might not possess anything at all. What is known however, is that the Romans compiled and kept route maps, known as *itineraries*, linking routes between towns and cities and other prominent landmarks (such as villas, temples, baths, but also mountain passes) and, importantly, showing distance information. These itineraries were akin to modern-day urban transit maps, just like those above carriage doors in underground trains: the maps show only vital information, removing all unnecessary topographic detail. They are devoid of any scale or compass direction, so tell us almost nothing about the geography, retaining only the relationship between points along the route depicted as straight lines.

The *Tabula Peutingeriana*, or Peutinger Table, was one such itinerary.[8] The map, drawn on a parchment scroll 34 centimetres in height and almost 7 metres in length, depicts the extent of the Roman road network at the height of the Empire around 375 CE, containing detailed information about points along the way. In total there are around 3,600 toponyms recorded on the map which stretches from the British Isles in the west to the Indian sub-continent in the east. To the modern eye however, this map of Eurasia is scarcely recognisable: the profiles with which we are familiar on the maps of today have been flattened, stretched and kneaded in such a way that they fit onto a format that was both manageable (for carrying around in rolled-up form) and legible on their chosen medium (papyrus rolls). Key topographical elements, such as mountains (crenulated markings), rivers (wavy lines) and forests (lines of trees), feature sporadically as other geographical reference points.

Once your eyes have adjusted to the unusual topological representation of the map and you know where to look, it's possible, amongst other things, to pick out the key Roman routes across the Alps. For example, from Turin - marked on the map as *Avgvsta Tavrinorvm* - the road to the summit of the Great Saint Bernard (*Insvmmo Pennino*) goes by way of Ivrea (*Eporedia*), Aosta (*Avgvsta Pretoria*) and Saint-Rhémy-en-Bosses (*Evdracinvm*), and then across the mountains (brown crenulations) by way of the pass to Martigny (*Octodvro*). Other Alpine passes are also marked on the Peutinger Table: *In Alpe Cottia* (Montgènevre); *Cvnvavrev* (Stelvio, an alternative route for the Septimer); and *Matreio* (Brenner).

The Peutinger Table is a medieval facsimile which was most likely copied from a Carolingian version, which in turn is assumed to be a copy of a third or fourth century Roman original. It is said to be loosely based on an earlier map first commissioned by the Roman general Agrippa in the first century and later engraved into marble near the Forum in Rome. The derived medieval copy came into the hands of an Augsburg-based scholar, Konrad Peutinger (1465-1547) who recognised that it was an important road map of the Roman Empire, hence its name. It is now kept in the Austrian National Library in Vienna.

By no stretch of the imagination can the Peutinger Table be considered a mountain map, let alone a topographical map, given its warped perspective and scant reference to natural features. Instead, the map, which shows the vast extent of the Roman road network and settlements, is an implicit expression of the Romans' world view, attempting – as the construction of the *Tropaeum Alpium* in Monaco did several centuries previously – to demonstrate their power and control.

So can the map tell us anything about the mountains, or more specifically the Alps? Well, yes. The mountain passes, clearly identifiable on the map, were the corridors by which access to their colonies on the other side was facilitated. By establishing and holding on to these lines of communication through a system of fortifications, encampments and tollhouses, they were able to oil the wheels for the large-scale movement of troops and goods in either direction, thus by-passing longer and more complicated routes. However, the map marks nothing more than a rudimentary starting point and it would be many more centuries before anyone would be able to make any sense of the mountains, at least in cartographic form.

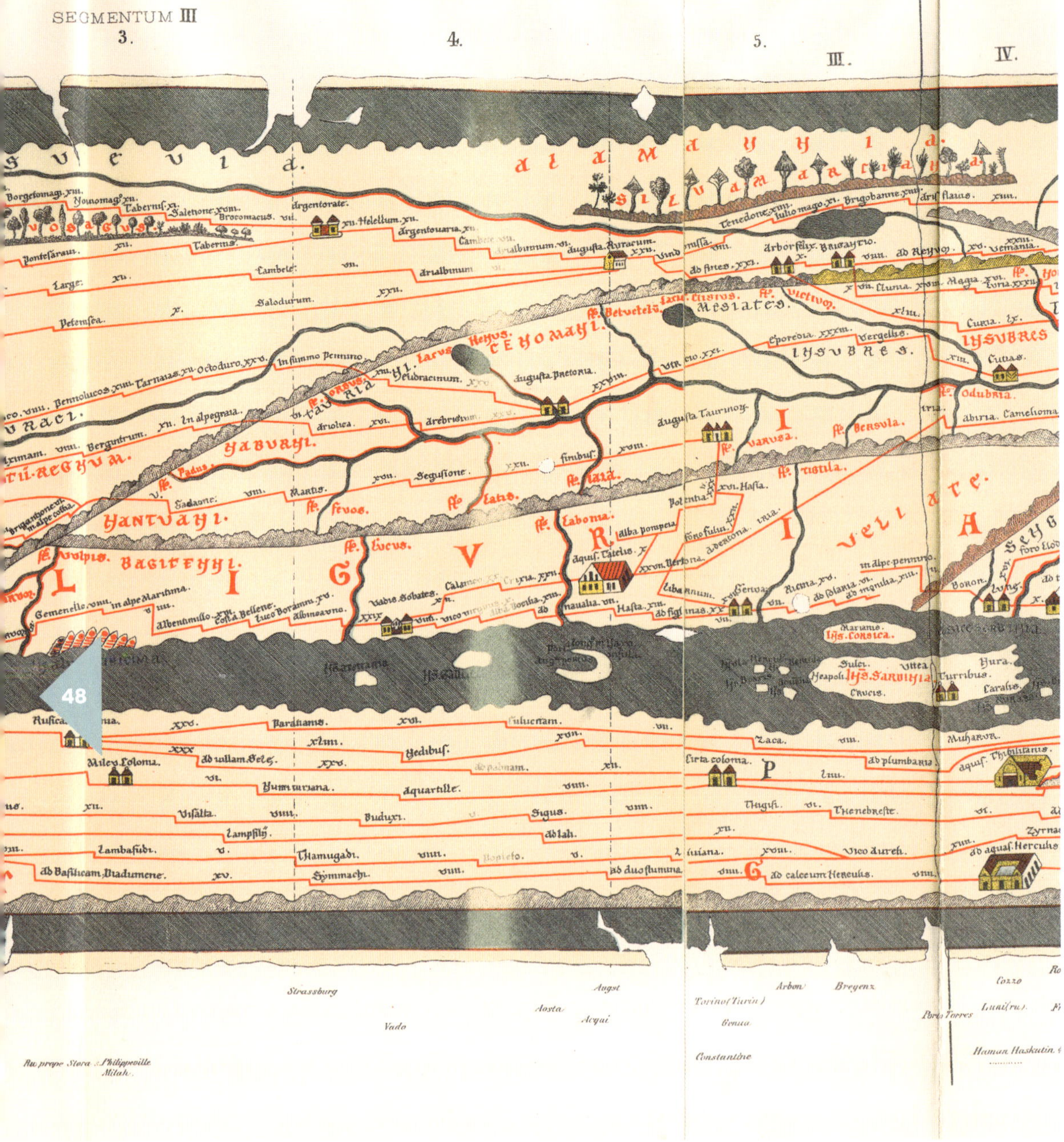

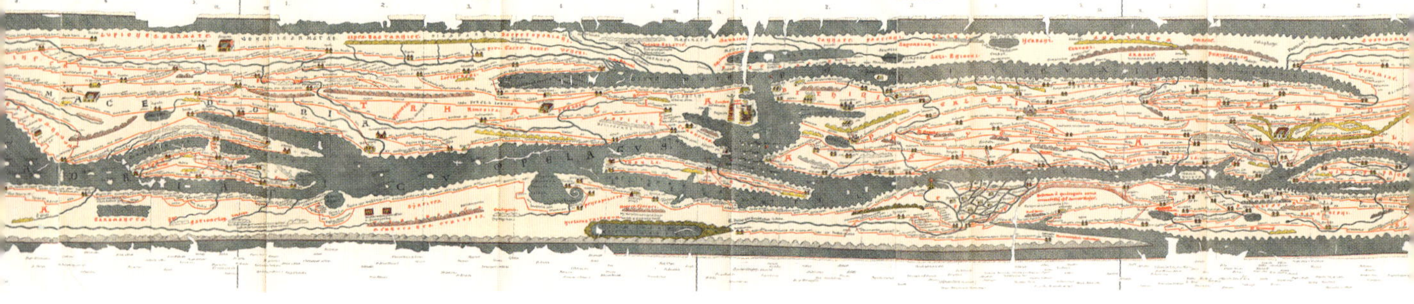

The *Tabula Peutingeriana* measures 34 centimetres by almost 7 metres.

Endnotes

1 Sometimes known as the Trophy of Augustus. In ancient Greek and Roman times, a 'trophy' was a memorial to victory erected on a battlefield or in another public place. Remarkably, the trophy at La Turbie remained intact (with additional fortifications in the Middle Ages) until it was partially blown up by troops of Louis XIV in 1705. Parts of the edifice were rebuilt in the early 20th century. Despite the semi-ruinous condition of the monument, the scale and the grandeur of the trophy can still be appreciated today.
2 Normally, such trophies were dedicated to divinities. In effect, the trophy at La Turbie deified Emperor Augustus.
3 In French, *Le col du Grand-Saint-Bernard.*
4 In pre-Roman times, the pass and the surrounding mountains went by the name of *Poeninus* or *Poenini*, supposedly the name of a mountain god who presided over the crossing.
5 The accompanying photograph was taken in early October.
6 The Little Saint Bernard pass (2,188m) – or Col du Petit-Saint-Bernard is at the western end of the Aosta valley on the French-Italy border.
7 At a macro-level, maps produced by the Romans served a geopolitical purpose, such as Agrippa's world map. At a micro level, land registration was underpinned by maps based on a system of *centuriation.*
8 The 13th century replica of the original *Tabula Peutingeriana* is kept in the Österreichische Nationalbibliothek in Vienna and is listed on Unesco's Memory of the World (MoW) Register.

Detail from the *Tabula Peutingeriana* showing the Western Alps on the right. It is possible to make out Turin *(Avgvsta Tavrinorvm)*, the Great Saint Bernard *(Insvmmo Pennino)* and Aosta *(Avgvsta Pretoria)*.

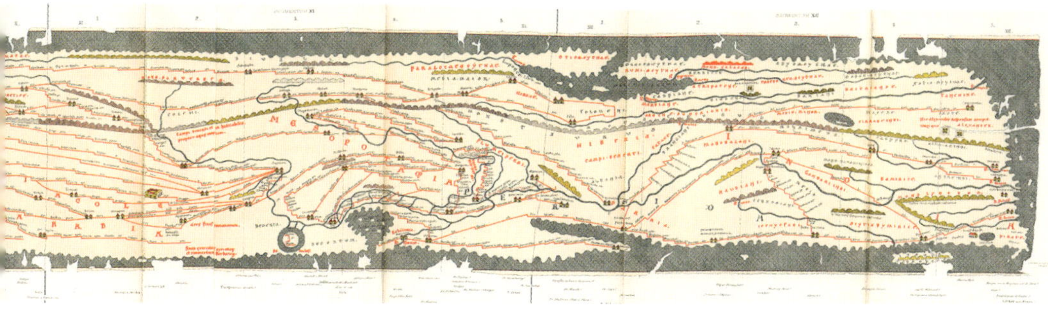

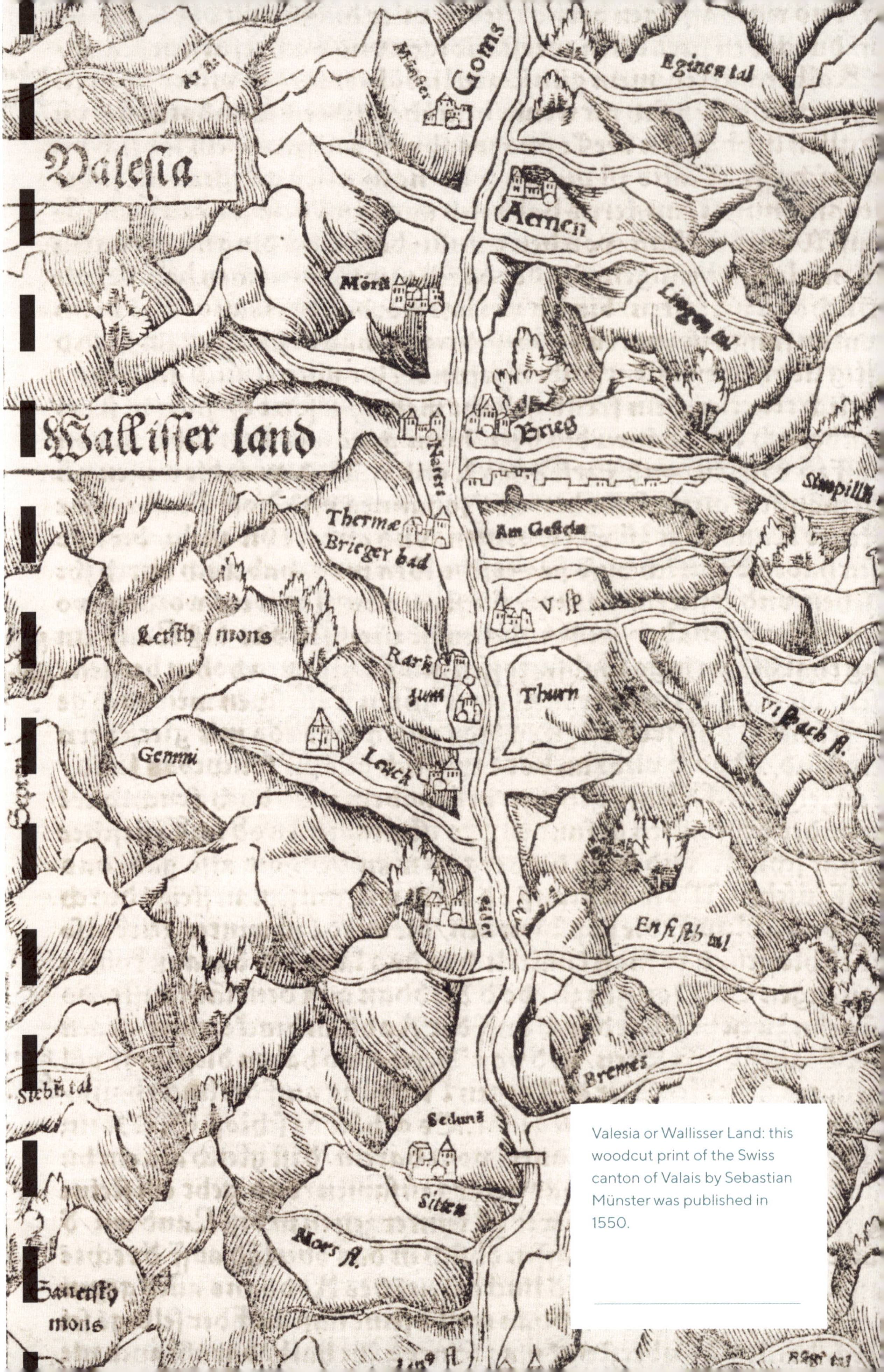

Valesia or Wallisser Land: this woodcut print of the Swiss canton of Valais by Sebastian Münster was published in 1550.

Early mapping in the Alps

Early mountain maps were often created on the basis of travellers' tales and a vivid imagination rather than through rational observation

Mountains were merely hinted at, or included as watershed symbols or as decorative filling for empty spaces – Edward Imhof

In the medieval world the development of cartography was held back by shortcomings in scientific and technological know-how. Towards the end of the Middle Ages, the printing press, advances in measuring instrumentation and – not insignificantly – the rediscovery of ancient texts on cartography, brought an end to this state of stagnation. Until then, maps had been drawn by hand on paper or parchment and so lacked consistency in quality and the potential for widespread distribution. With the Renaissance however, there came a cultural and technological transformation in numerous areas of intellectual life, not least in art and architecture, politics, religion, literature and, of course, science and technology. One aspect of the latter was the gradual proliferation of printed matter, including maps.

This cultural revolution also coincided with the opening up of worlds beyond Europe and it prompted a growing need for maps and navigational charts. Exploration was driven by an ambition for power and domination, both economic and military. Yet interest in mountains, glaciers and uninhabitable regions – because of their lack of economic value – would remain muted, and there was little incentive for producing detailed maps of mountain regions. Maps of mountains that existed focused largely on major lines of communication along valley floors and across the main passes: the mountains that lay in between were indeterminate barriers, filled in casually and unsystematically.

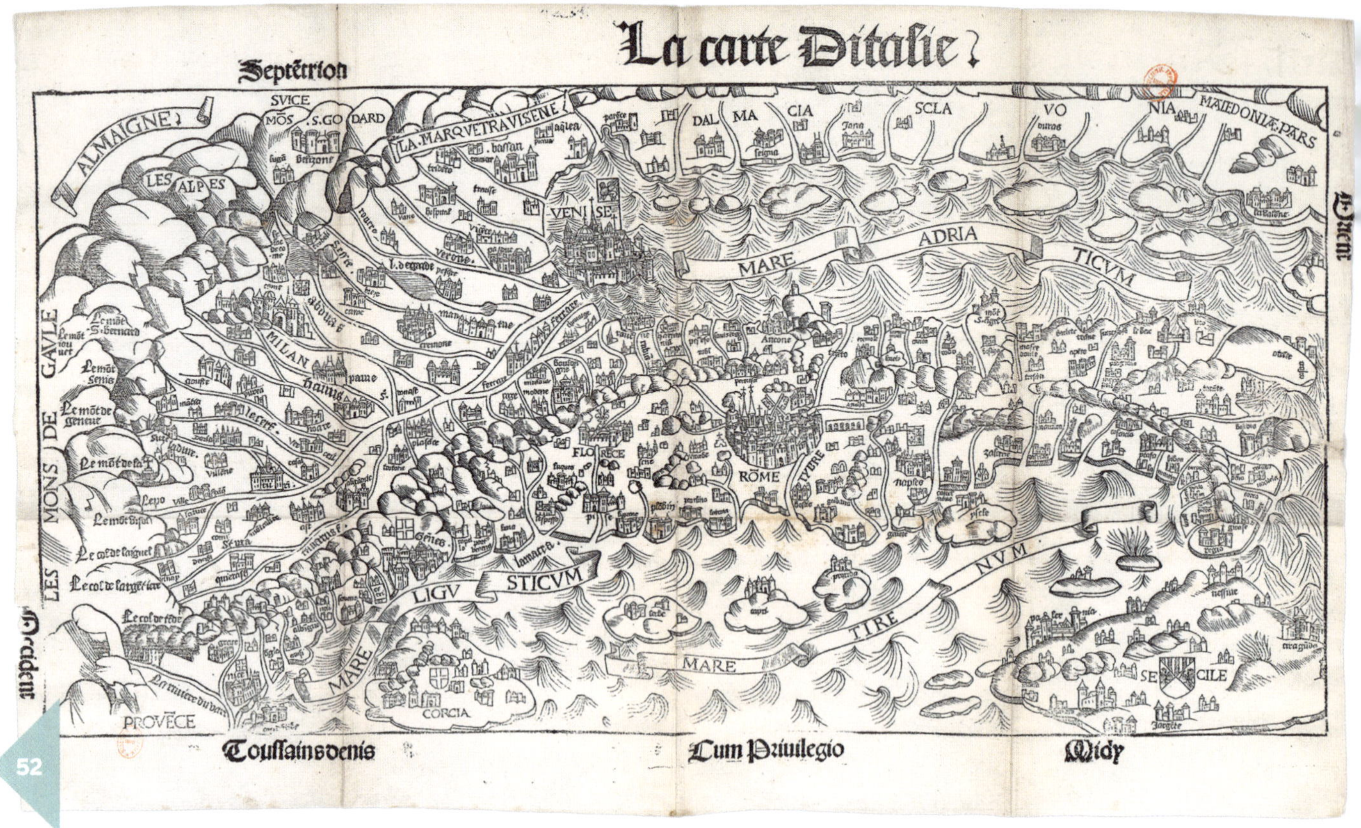

A 1515 map of Italy by Jacques Signot in which the Alps are portrayed as a largely smooth and bulbous series of rocks.

In the main, early Alpine maps were depicted pictorially and embellished in the style of the individual cartographer: uniformity and standardisation in all aspects of cartography were still a distant dream. Towns and settlements were often depicted by a church symbol or a group of houses, possibly accompanied by a coat of arms. Rivers were rendered out of scale to their true proportions, with glaciers completely absent. Depending on the cartographer's technique or whims, visualisation of the mountains might take on a horizontal or vertical perspective. An example of the former was to represent mountains as molehills or cones, which were often arranged in rows to symbolise a range of hills.

From a contemporary perspective, the impediments that topographic cartography faced in the transition to a modern, more systematic and rational-thinking world were twofold. First, how was it possible to measure more accurately the geographical relationships between natural and artificial features of the landscape, such as rivers, relief, roads and towns? And secondly, how could these relationships, which existed in three-dimensional form, be more faithfully depicted in a two-dimensional diagrammatic and readily readable setting? The two – surveying and cartographic design – worked in parallel. Only when instrumentation became standardised and there was some kind of conformity in cartographic drawing did maps begin to take on a form that is recognisable with (topographic) maps of today. Needless to say, mountain regions presented their own unique obstacle to mapmakers, not least their remoteness and the sudden and extreme changes in relief, the latter which did not lend itself easily to representation on a flat piece of paper.

Whether it was in the lowlands or the mountains, a map's accuracy depended on the reliability of angle and distance-measuring instrumentation. Although architects and surveyors in the Middle Ages would have been aware of arithmetic and geometric principles, for example, in the construction of the great cathedrals, these were not usually applied to mapping until the Renaissance. Breakthroughs included the application of trigonometry and astronomical observation. Over time, quadrants, sextants, telescopes, plane tables and rudimentary theodolites would be used to measure angles (whether celestial or terrestrial) and ropes, rods, chains and surveyor's wheels to define distances. Only then did cartography slowly become a science.

Setting aside their lack of practical economic value, it should not be underestimated how difficult objective surveying in the mountains would have been without accurate equipment and adequate supplies. Surveyors venturing into the high Alps had to endure considerable discomfort, even when new and more systematic techniques of mapping began to evolve. Maps once produced on the basis of travellers' tales and a vivid imagination were now being created through rational observation using instrumentation. The use of baselines and the calculation of angles between landmarks ultimately enabled surveyors to penetrate mountainous regions on the fringes of their maps, but until equipment became more lightweight and portable, it was always going to be a trying experience in more extreme terrain.

Attempts to translate mountainous relief accurately into a two-dimensional planimetric form came slowly and only later did maps start to appear which included some shading to provide an indication of a mountain's inclination and shape. Improvements in printmaking media also enabled a step-up in cartography: woodcuts were superseded by the superior copper engraving which allowed the representation of more topographic detail on a map.

As for the high mountains, they would remain largely untrodden by mapmakers and mountaineers for some considerable time to come. To Renaissance people, the Alps were still the mountains through which such legendary figures as Caesar and Hannibal had passed, and they were viewed in much the same manner by the first Enlightenment cartographers. But over time, for both military and economic reasons, rulers of territories which abutted the Alps would start to take a much keener interest in delineating their sovereign borders. Tentative at first, this desire inevitably led to the penetration of the high mountains. Over the next centuries, as surveying and cartographic skills improved, and as strategic interests and tourism grew, this progressive encroachment into the mountains helped towards filling in the gaps.

Swiss-style
Snapshots from a nation's cartographic past

There are surely few mapping agencies worldwide that can rival the superlative attention to detail given by swisstopo, Switzerland's national mapping agency. In terms of innovation in mountain cartography, the Swiss enjoy a formidable reputation. Clearly, a country 70 percent of whose land surface is classified as being mountainous[1] was always going to have a head start in this respect, but Switzerland's rich tradition of mapmaking cannot be solely put down to its rugged terrain. In the late medieval period, the country's proximity to early centres of printing and publishing and the burgeoning of universities in German-speaking lands were an important stimulus, not to mention the nation's fierce spirit of independence.

Located on a bend in the Rhine in the north-west corner of present-day Switzerland, Basel was one of the cities that flourished culturally and economically during the late Middle Ages, benefiting from the synergies of being close to other centres of trade and learning along the Rhine valley, such as Freiburg, Strasbourg, Heidelberg and Mainz. By the second half of the 15th century, the city had its own university and had already established a name for printing and publishing. In 1501, the city joined the Swiss Confederation. It soon became known as a centre of mapmaking, with a number of ground-breaking innovators of Swiss cartography being associated with the city, such as Konrad Türst, Sebastian Münster and Aegidius Tschudi.

The astronomers and geodesists of this period were initially attracted to Ptolemy and the reproduction of the *tabulae antiquae*, forerunners of the world atlas, but there was also a developing interest in regional maps, and this was particularly so in Switzerland. At the time, the country was a patchwork of independent states (or cantons), intensely protective of their own identities, but nevertheless existing under the umbrella of a Swiss Federal Diet.[2] Its cartographic development and traditions must therefore be seen against this historical and political backdrop.

One of the earliest surviving maps of Switzerland is an exquisitely hand-drawn map by Konrad Türst (1451-1503). Türst graduated from the University of Basel in 1470 and after further studies in Ingolstadt and Pavia, became a physician in Zürich. He is best known however, for his pen and ink drawn map on parchment of the early Swiss Confederacy, drawn at an approximate scale of 1:500,000. Two maps survive: one in Zürich and another in Vienna (both of which have accompanying descriptions in German and Latin respectively). From the bottom, the map has a north-west to south-east perspective and depicts mountains, woods, settlements, rivers and lakes, but curiously no roads. Latitudes and longitudes are marked along its edges.[3] Mountains are painted in green with the steeper rock faces of the Bernese Oberland in brown. With the Po Valley, beyond the southern confines of the Alps, rendered indistinctly in the background, it is possible to pick out the Saint Gotthard (Adula) and Great Saint Bernard (S. Bernhartz) passes. The overall impression is of a visually pleasing and uncluttered map drawn by a talented artist and this combination of the technical and the artistic has been a defining feature of Swiss cartography right up to the present day.

One of the earliest surviving maps of Switzerland is an exquisitely hand-drawn map by Konrad Türst dating from around 1495. The map is a combination of the technical and the artistic, a defining feature of Swiss cartography right up to the present day.

Cartographers Münster and Tschudi were contemporaries of each other whose paths probably first crossed in Basel. Sebastian Münster (1488-1552) was born in Ingelheim, near Mainz. Although he studied theology and Hebrew, the latter which he went on to teach during much of his later life at the University of Basel, he is first and foremost known as a leading light in early German geography and cartography. Münster, a cosmographer who held a wider world view of mapmaking, took an interest in editions of Ptolemy's *Geographia* and eventually brought out his own copy in 1540. But he also dabbled in hands-on cartography, designing instrumentation and producing a map of the Heidelberg region in Germany, which was based partly on actual surveying principles.

The mapmaking career of Aegidius Tschudi (1505-72), on the other hand, focused almost entirely on his native country of Switzerland. He was born into a well-to-do family in the town of Glarus, south-east of Zürich, and his formative years were spent under the tuition of Glaraenus[4], amongst other things an expert in globemaking from whom he must have developed a passion for geography.

Tschudi later occupied various positions in officialdom in Switzerland and travelled widely for his work, including the crossing of many Alpine passes. He accumulated vast amounts of information about the topography and history of the country. This included the drawing of maps, and by 1528 he had amassed enough detail in manuscript form to proceed with publication of a printed map of Switzerland. Tschudi apparently hesitated with its publication, possibly in anticipation of further improvements, but also because he may have been otherwise occupied.

At some point, the manuscript maps fell into the hands of his tutor Glaraenus and it is said that when Sebastian Münster got wind of this, he asked to borrow them for a viewing and had them secretly copied. Münster went on to publish his own map in printed form in 1538, unbeknown to its original author. In a letter, Tschudi later complained that it was 'against my will, wish, and order, behind my back.' But it is also known that the two cartographers were otherwise great friends, so the alleged story of Münster's underhand dealings may have been stretched to make an interesting cartographic aside.

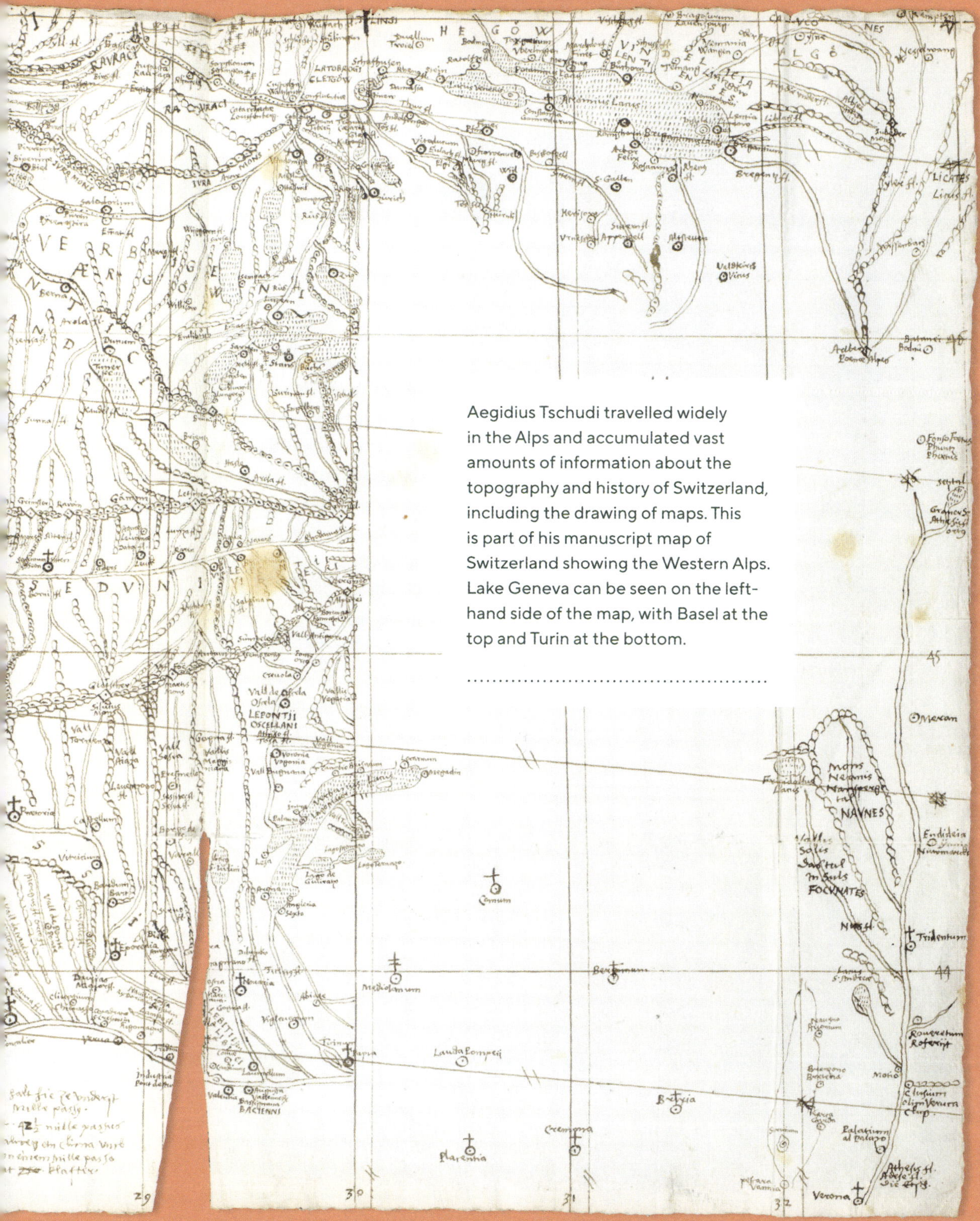

Aegidius Tschudi travelled widely in the Alps and accumulated vast amounts of information about the topography and history of Switzerland, including the drawing of maps. This is part of his manuscript map of Switzerland showing the Western Alps. Lake Geneva can be seen on the left-hand side of the map, with Basel at the top and Turin at the bottom.

Sebastian Münster's 1545 map of Valais, drawn by Johannes Schalbetter, with south at the top. In the centre foreground is the town of Sitten (Sion) and in the top left is Mons Silvius, believed to be the Matterhorn.

Tschudi's original 1538 map of Switzerland, based on his manuscript drawings, has been lost and the earliest surviving copy held by the University of Basel is from 1560. It is the first map, possibly the first true mountain map, that charts the Alpine region with any accuracy and contains toponyms found in later maps of the country. Like many early maps of Switzerland, it is oriented with the south at the top.

Whilst cartographers elsewhere were churning out world maps, the Swiss – landlocked and without access to a global trading network – continued producing regional maps of their own backyard. The man credited with producing the first 'atlas' of the newly formed federation was likewise a German who had made his home in Switzerland. Johann (or Johannes) Stumpf (1500–c.1578) was born in Bruchsal near Karlsruhe. After studying in Strasbourg and Heidelberg, he joined the priesthood in 1522 and was assigned to a parish in Bubikon, a small settlement near Zurich, where he remained until 1543. His ecclesiastical activities allowed him to travel around Switzerland and he developed an acute sense of the country's history and natural beauty.

During his time in Bubikon, Stumpf started to plan a comprehensive history and topographical description of Switzerland in the form of a Chronicle, and spent many years writing, collating, organising, transcribing and editing materials of the country's geography and history. The Chronicle, which went under the title of *Gemeiner loblicher Eydgnoschafft Stetten, Landen und Voelckeren Chronick wirdiger thaaten Beschreybung*, was published as a 934-page folio in Zurich in 1548. Thirteen maps from

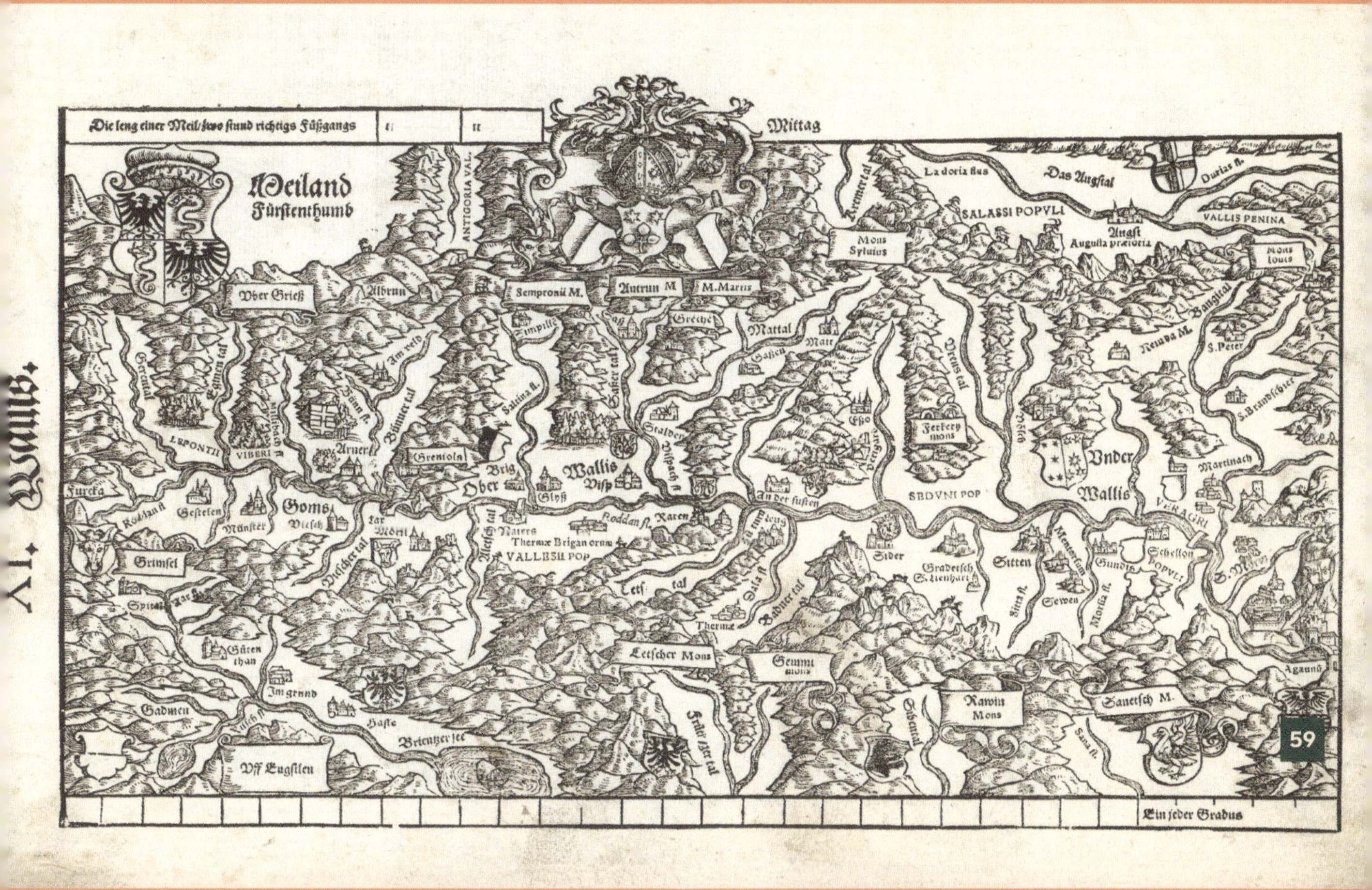

Map of Valais by Johann Stumpf with south at the top. In 1552, Stumpf published thirteen maps, or Landtafeln, depicting regions of the Swiss Confederation. It is considered to be one of the first attempts to create a uniform collection of maps, or atlas.

Stumpf's Chronicle were published separately in 1552 as *Landtafeln* (literally, 'land tables') and they represented a first attempt to create a collection of maps that were uniform in format and presentation. As such, they were considered to be the first true atlas of the Swiss confederation.

Endnotes

1 By the Swiss government.
2 This diet – or legislative assembly – existed in several forms until the formation of the Swiss federal state in 1848.
3 It is known that Türst made astronomical observations. The latitude of Bern for example, is six minutes out, but because his map is shown to be distorted against a modern-day grid of longitude and latitude, they were probably at best guesstimates and based on distances between settlements taken from hearsay or personal experience.
4 Henricus Glaraenus (1488-1563), also referred to as Heinrich Glarean or Heinrich Loriti, was a Swiss humanist, most widely known for his interest in music theory and poetry.

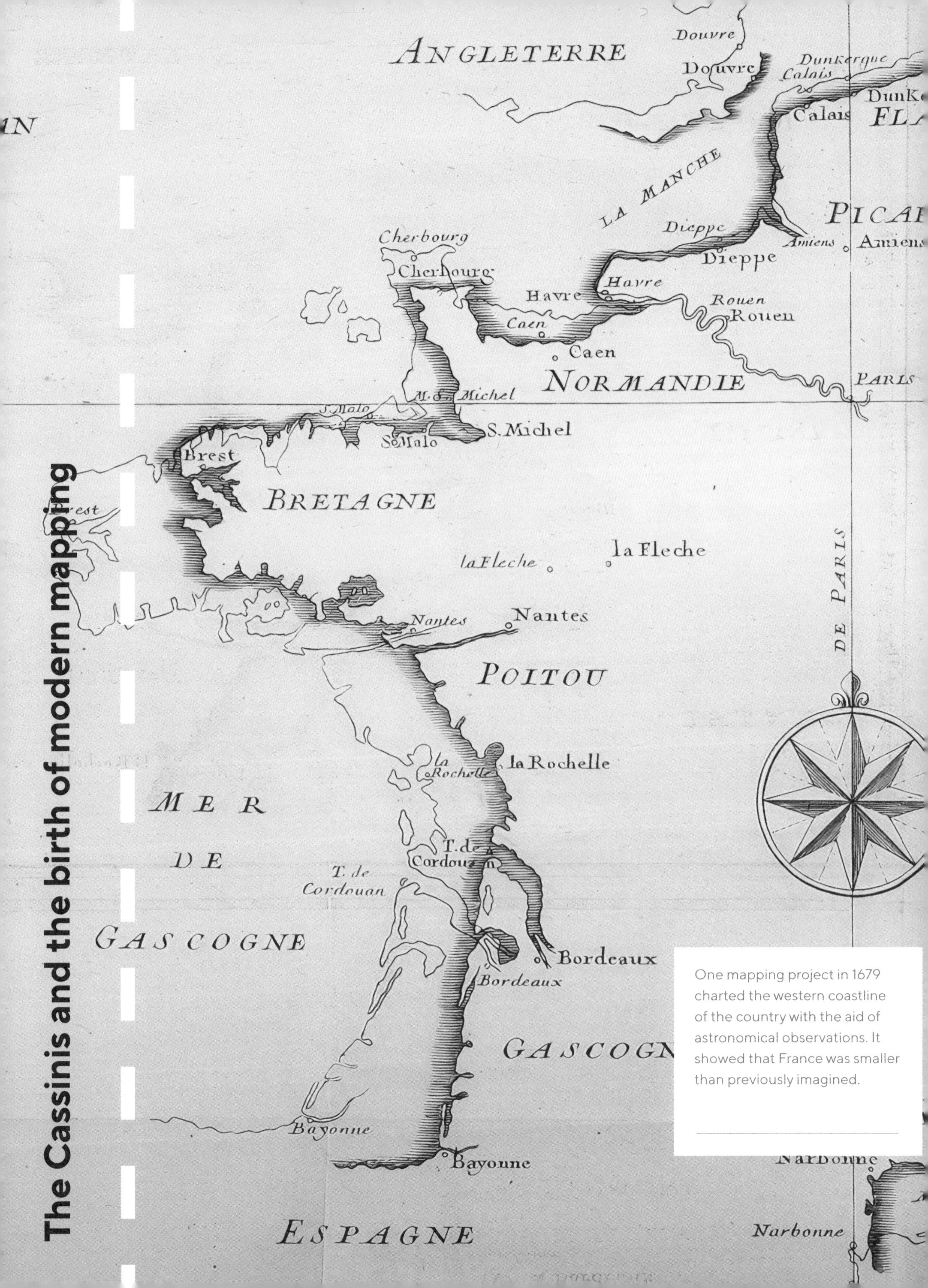

One mapping project in 1679 charted the western coastline of the country with the aid of astronomical observations. It showed that France was smaller than previously imagined.

The Cassinis and the birth of modern mapping

If any name is synonymous with the birth of modern mapping it is Cassini: not so much a name, but a dynasty of astronomers and surveyors

Giovanni Domenico Cassini (1625-1712), an accomplished Italian astronomer, was brought to France in 1668 by King Louis XIV to help implement a prestigious plan to put the country at the forefront of modern science. One of Cassini's tasks as director of the royal observatory in Paris was to produce an accurate map of the country. This would be based on astronomical and trigonometric principles. The reasons for making such a map were diverse, but it was mainly for military, administrative and economic purposes.

One project, commencing between 1668 and 1670, was to measure an arc of a meridian from north to south between Paris and Amiens, the ultimate objective of which was to calculate the diameter of the Earth by measuring one degree of latitude. A later project in 1679 mapped the western coastline of the country with the aid of astronomical observations tracking the eclipses of Jupiter's moons. The map, which was completed in 1682, was a surprise: its longitudinal accuracy made France smaller than previously imagined. Louis reportedly quipped that he had been robbed of his territory.

When Giovanni died in 1712, he was succeeded as director of the observatory by his son Jacques Cassini (1677-1756). The so-called 'Cassini II'[1] was less of a geographer and more of an astronomer, signalling a shift from cartography to geodesy, that is, measuring the shape and the size of the earth. To do this, in the 1730s, France sent expeditionary teams to Lapland and Peru[2] respectively to calculate the length of a meridian arc in the Tropics and the Arctic. Through a combination of astronomy and triangulation, the expeditions confirmed that the Earth was an oblate spheroid (i.e. flattened at the poles) rather than a prolate one. In the meantime, astronomy and triangulation had become established as the two pillars on which modern cartography was founded.

This scientific pursuit was noble, but in the 1730s calls were growing for more accurate maps of France itself. A baseline was first assiduously staked out near Paris and teams of surveyors and engineers were then sent into the field to carry out triangulation work. By 1744, 800 principal triangles and 18 more baselines had been measured. This network of triangles provided the basis on which the topographical details would be filled in over later decades under César François Cassini de Thury (1714-1784), otherwise known as Cassini III, the son of Jacques. It was a mammoth undertaking – one which would involve the mapping of the country's mountains, rivers, towns and villages – and which met with considerable logistical and financial obstacles along the way. What had been forecast to take 18 years took the best part of half a century and was ultimately completed by Jean-Dominique Cassini (1748-1845), known as Cassini IV. The result was one hundred and eighty uniform maps of the country produced at a scale of 1:86,400.[3]

It was in the late 1770s that the last of these maps were published, more specifically, those of the French Alps. The mountains provided surveyors with unfamiliar terrain and they were ill-equipped to give the same level of topographic detail as in the lowland regions, preferring to survey from the more comfortable fringes of the region, rather than from the nigh inaccessible peaks at the centre in the mountain ranges.

Nevertheless, at the time, the maps were unrivalled in their cartographic precision, their depiction of relief and their use of signs, symbols and lettering. They would become a benchmark for future mapping and provide the inspiration for national surveys in the rest of Europe.

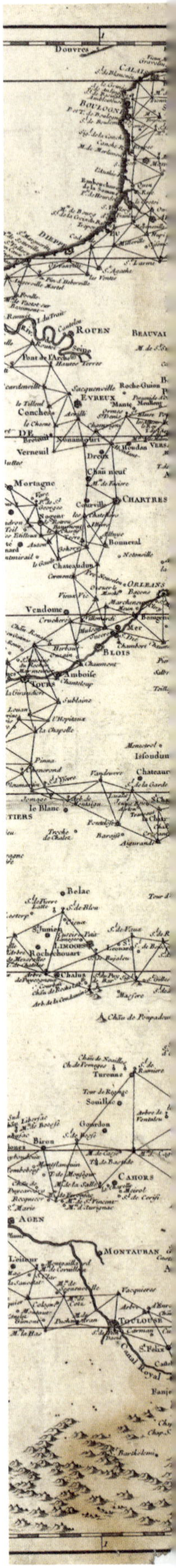

Endnotes

1 Regnal numbers are often used to distinguish between the different Cassini generations.

2 See chapter 6.

3 This scale was based on standard units of measurement used in France at the time. 1 *toise* (roughly equivalent to 1.94 metres) was divided into 864 *lignes*. The scale of 1:86,400 therefore represented 1 *ligne* to 100 *toises*.

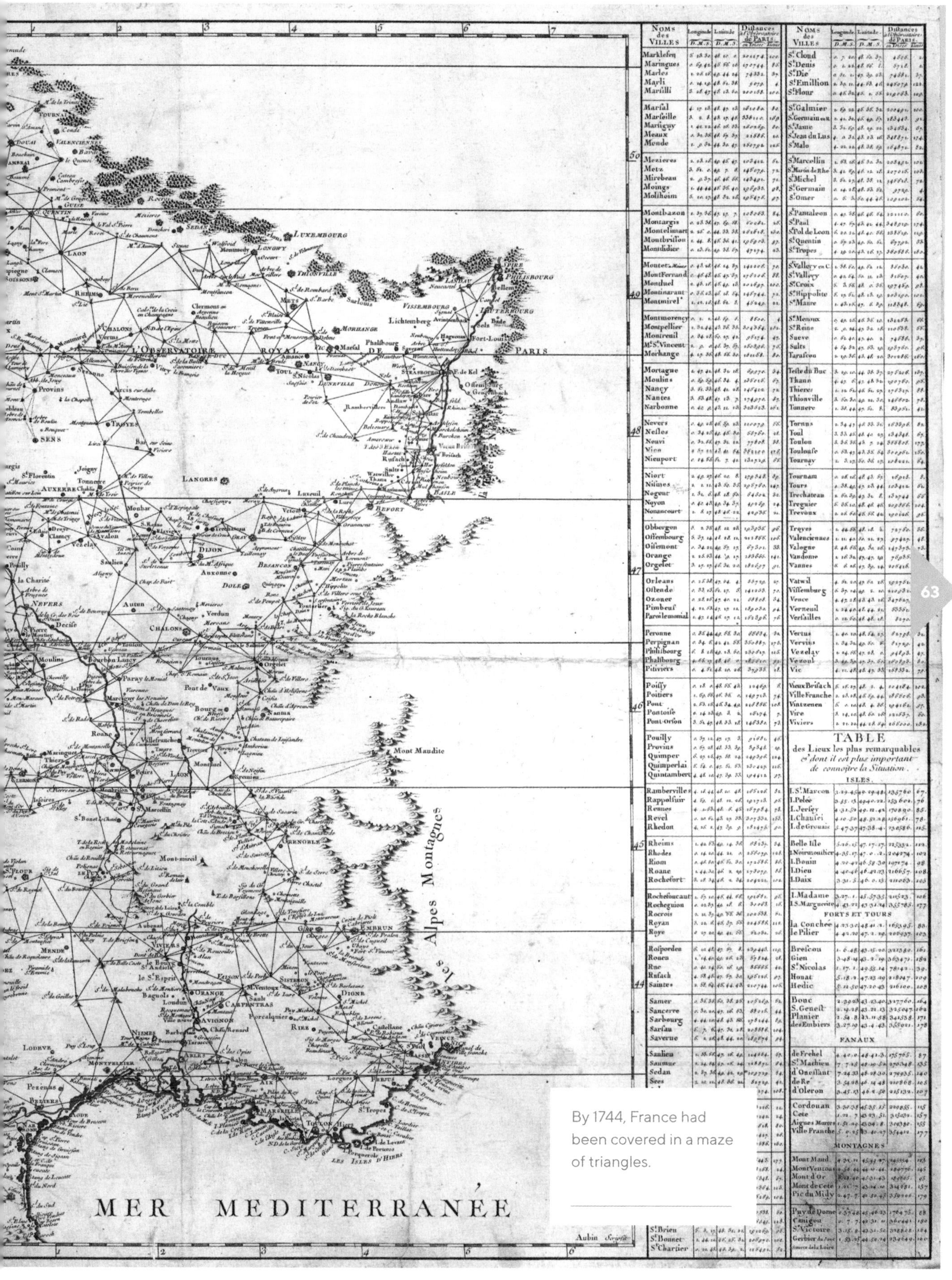

By 1744, France had been covered in a maze of triangles.

The Briançon sheet of the Cassini map, published in 1779.

F. 124
PIEMONT
le Lauzet
le Monetier
les Guiberts
la Salle
Chantemerle
St. Chaffre
Puy St. Pierre
Puy St. André
Servieres
les Vigneaux
Queyrieres
Champcella
Reotier
Guillestre
St. Clement
Chateauroux
Vars
Orcie
Echelle de dix mille Toises

Swiss-style
Snapshots from a nation's cartographic past

Swiss mapmaking underwent a quantum leap forward in the 17th century when, for the first time, planimetric maps started to appear. This type of map depicts topography and relief from above in a 'plan view', rather than from the side in an 'elevation view': the verticality of hills and mountains are, as it were, reduced to a two-dimensional form (instead of being drawn as cones or molehills).

The individual responsible for this advance was Hans Conrad Gyger (1599–1674). Gyger was born in Zurich and followed in his father's footsteps as a glass-painter. He is said to have developed an interest in mapmaking from his teacher Josias Murer, who was also a cartographic draughtsman. It was probably in 1620, when he was employed by the military authorities to draw a map of north-eastern Switzerland, that Gyger first established himself as a skilled cartographer.
At the time, the Swiss used signal towers across the country as part of a network of early-warning systems, but these individual towers were joined to each other by lines that formed part of a triangulation network. Spanning a period of over 38 years, operating under strict secrecy, Gyger created a series of 56 ink-drawn manuscript maps of the Zurich canton which ultimately served as a template for a much larger map of the whole canton produced in 1667 as an oil painting. Measuring about five square metres, the map presents the landscape from a vertical viewpoint using natural colours and shaded relief, its accuracy enhanced by the use of the leading-edge surveying techniques of the day. According to some, it is the earliest example of a relief map in existence, but ironically failed to have any influence on contemporary mapmakers as it was kept hidden away as a military secret.[1]

Further evidence of Switzerland's growing reputation for relief depiction was provided by Franz Ludwig Pfyffer (1716-1802) towards the end of the 18th century. Born in Lucerne, Pfyffer followed the example of his forefathers and embarked on a career in the French army serving with a legion of Swiss Guards at the court of King Louis XIV. Amongst other things, the guards were responsible for protecting French royal households. From an early stage in his career, Pfyffer would have become familiar with terrain models of fortresses created by the armed forces,[2] and during field campaigns, he would have picked up knowledge of surveying from artillery engineers.

At some point during his many visits to his home city of Lucerne, Pfyffer conceived the idea of making a relief model of his native region taking in the mountainous terrain of the *Urschweiz* in central Switzerland, made up of the original cantons of Uri, Schwyz and Unterwalden. After 24 years in the making, the model was finally completed in 1786. Measuring 3.9 by 6.6 metres, the model depicts over 4,000 square kilometres of topography - about one-tenth the present size of Switzerland - at a scale of 1:12,000 for the horizontal and 1:10,000 for the vertical. He used several materials in its making, like bricks, clay, charcoal, putty, pitch, cardboard and wax.

One of 56 manuscript sheets of Gyger's map of the Zurich canton, showing part of Lake Zurich, with the east at the top

Gyger's 1667 oil-painted map of the Zurich canton was 38 years in the making, but its military importance meant the map was kept hidden away for a long period of time.

Over the course of its evolution, Pfyffer used the summer months to travel the length and breadth of the surroundings, collecting detailed information on the topography which was encapsulated in 73 sketches. Lines radiating out from each of the locations he chose indicate the use of triangulation techniques that were now becoming commonplace among contemporary European mapmakers. Pfyffer's sketches were marked using different water colours to indicate landscape features, such as permanent snow and glaciers, forests, pastures, bodies of water, settlements, etc., which matched the colourings on the relief model.

After its completion, it went on to become an object of wonder in Switzerland, attracting visitors from far and wide, including such notables as Goethe. Pfyffer's relief model also served as a basis for panoramic views and maps that followed, and its detailed terrain information was also said to have helped military strategy during the Napoleonic Wars.[3]

Print of Pfyffer taking observations on Mount Pilatus overlooking Lucerne.

Endnotes

1 The original map is currently kept at the *Staatsarchiv des Kantons Zürich*.
2 The tradition of building three-dimensional models – or *plans-relief* – of fortified towns in France dates back to the mid-17th century. A collection of these is held at the *Musée des Plans-Reliefs*, a museum of military models located at the Hôtel des Invalides in Paris.
3 Since 1873, the relief model has been on display at the *Gletschergarten* museum in Lucerne.

Christian van Mechel's 1786 panorama of the Lucerne canton looking south-east was drawn on the basis of Pfyffer's relief model.

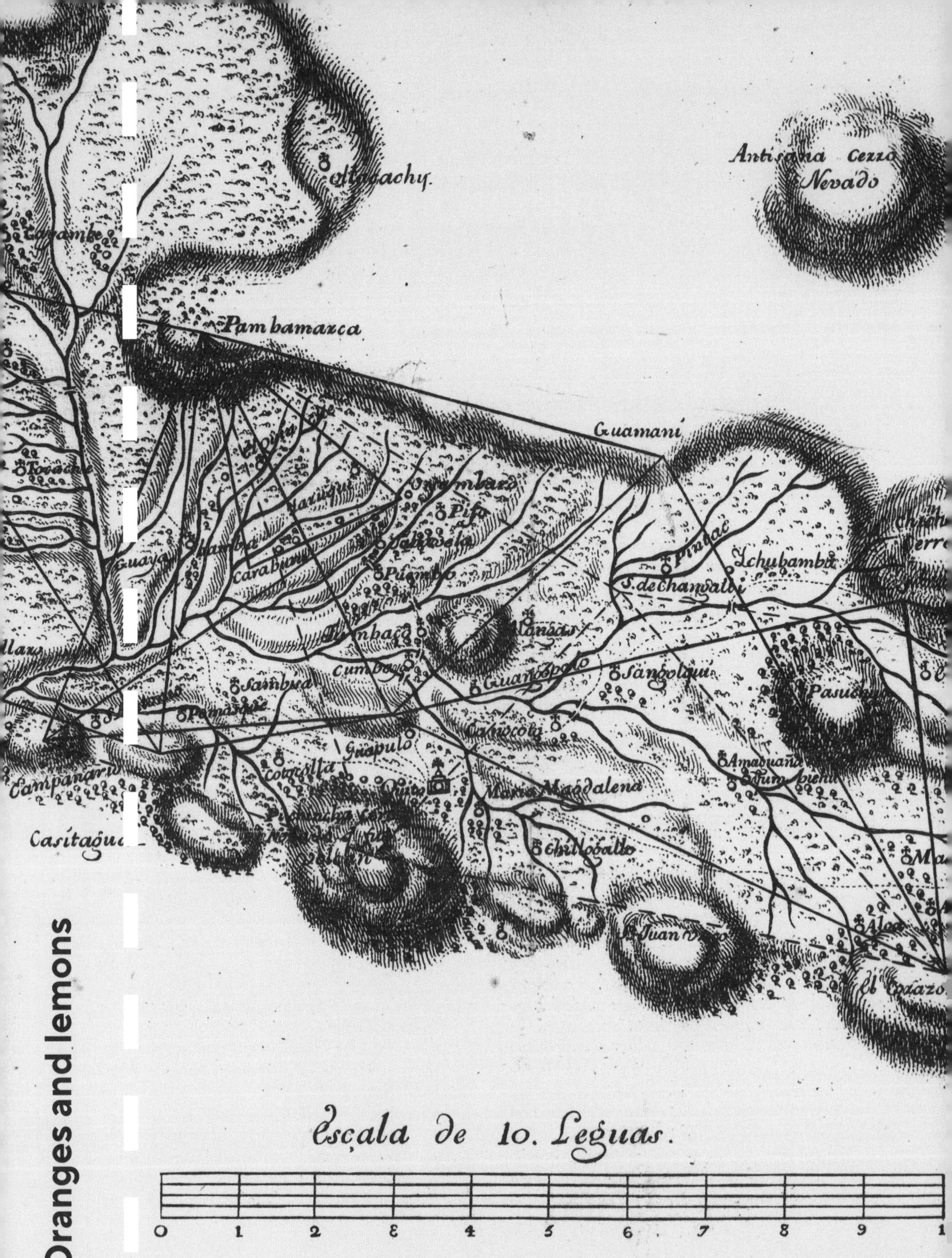
Antisana Cerro
Nevado
Pambamarca
Guamani
Ichubamba
Sangolqui
Guapulo
Maria Magdalena
Campanario
Casitagua
Escala de 10. Leguas.
0 1 2 3 4 5 6 7 8 9

Detail of the map of the meridian arc by Jorge Juan and Antonio de Ulloa.

Chapter 6

Oranges and lemons

In 1735 a team of French scientists journeyed to the Equatorial Andes on an incredible odyssey to measure the shape and size of the earth

In the first half of the 18th century an argument raged between the Newtonians and the Cartesians as to whether the Earth was an orange or a lemon. Or rather: whether it was shaped like an orange or a lemon. Followers of Isaac Newton maintained that the centrifugal forces of a rotating spheroid would produce flattened poles and a bulging equator. Conversely, supporters of René Descartes' theory of vortices were of the view that the Earth's circumference was greater at the poles than on the equator. On the face of it, this problem might seem purely hypothetical, but the French Academy of Sciences believed it was critical to know the exact shape of the Earth for safer and more effective navigation.

To settle the debate, in 1735 and 1736, the Academy sent expeditions to the equator and the Arctic Circle respectively to calculate the length of one degree of latitude: if this were shorter at the equator, the Earth was oblate (e.g. an orange) and, if longer, it would be prolate (e.g. a lemon). The aim of each mission was to establish a chain of triangles along a meridian arc and to take astronomical observations at either end of this chain.

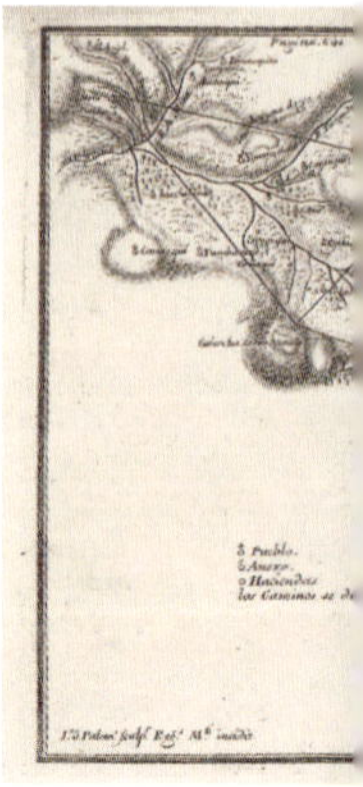

By the standards of the day, this was a formidable undertaking. In particular, the equatorial mission would prove challenging, involving travel through tropical rainforests, climbing peaks of a stature unparalleled in Europe, and other perils that included deadly disease and mistrustful locals. Whereas the mission to Lapland was completed in just over a year, the expedition to Peru was to last a good deal longer.

More than a year after leaving the French port of La Rochelle in May 1735, the mission finally assembled in Quito in the Spanish-controlled Viceroyalty of Peru from where geodetic operations would commence. The main protagonists of the campaign were the French geodesists and astronomers Charles Marie de La Condamine, Pierre Bouguer and Louis Godin, along with the Spanish naval lieutenants Jorge Juan y Santacilia and Antonio de Ulloa, appointed by the Spanish Crown to accompany the mission.

It was a disparate collection of individuals and after the year-long journey to South America, the group was beset by personality clashes and internal bickering. Nonetheless, by December 1736 they had set down a 12.23-kilometre baseline at Yaruqui, near Quito, around one-quarter of a degree south of the equator. Bouguer was optimistic enough to believe they would be finished with the triangulation by the end of 1737, but they had already run out of money and La Condamine was despatched on a 4,000-kilometre round trip to Lima, the capital of the Viceroyalty of Peru, to petition for new funding.

Their first experiences at the first triangulation station on Pichincha, starting in August 1737, were a portent of things to come. At 4,784 metres, the elevation of this volcano towering over the city of Quito was almost on a par with that of the yet-unclimbed Mont Blanc back in Europe. Ulloa suffered terribly from altitude sickness and collapsed on the ascent, his companions did not fare much better.

They built a makeshift hut at the top but were regularly pummelled by snow and hail, spent much of their time huddled together trying to keep warm and in the mornings had to be dug out by their local porters from the snow drifts that had accumulated outside overnight. Rarely through the clouds, were they able to take sightings of the markers at the two ends of the baseline far below. After 23 days, defeated, they decided to place a signal on the lower slopes of the mountain, from where it took a further three months to complete their angular measurements. Two and a half years after leaving France, they had completed their first triangle and there were another 30 to go.

Map of the meridian arc by Jorge Juan and Antonio de Ulloa, showing the triangulation across mountains in the Andes near the equator (with north oriented towards the left).

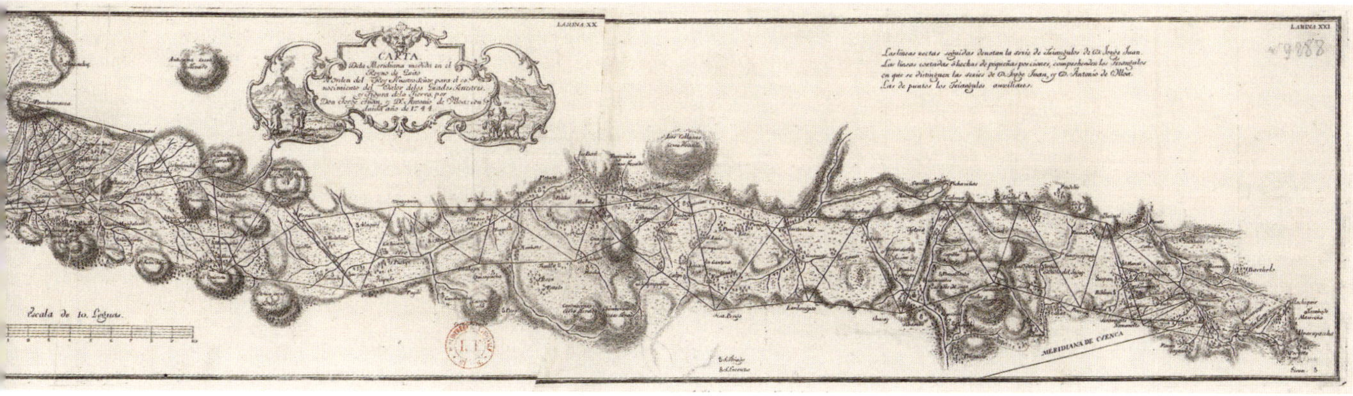

The eruption of Cotopaxi. In his journal, La Condamine reported that the flames rose at least 60 metres above the top of the mountain.

South of Quito, the mission planned to stake out 30 triangles as far south as Cuenca following a natural corridor flanked on either side by the western and eastern cordilleras of the Northern Andes, the peaks or the slopes of which could be put to use as triangulation stations. These included such leviathans as Cotopaxi (5,897 m) and Chimborazo (6,263 m). The French surveyors split into two groups, more by way of discord than desire, but this produced the benefit of measurements being doubled up, thus enabling comparison. As the scientists toiled their way south, they had to undergo similar privations as on Pichincha. Sometimes the wooden pyramids they built as markers would be stolen or get blown away in tempests. At other times, they would be stranded in raging storms for days on end without food or water.

Along the way, they became accomplished mountaineers however, and carried out scientific experiments, very much in the spirit of the Age of Enlightenment. Amongst the discoveries they made were rubber, platinum and the correct type of cinchona tree that produces quinine. They also witnessed the eruption of Cotopaxi in June 1742.

In May 1739, they finally arrived in Cuenca, where the two groups mapped the last of their triangles along a meridian arc of 345 kilometres. A control baseline was laid out at Tarqui by the two teams measuring 10.25 kilometres. Incredibly, in August, they discovered that the discrepancy of their calculations along the entirety of the trigonometric chain amounted to just 214 centimetres, a difference of 0.0006 percent.

Buoyed by their success, they set about the final step of their mission which was to determine the difference in latitude between the two ends of the chain through astronomical observation. Their goal was to fix the angle of the star Alnilam in the constellation of Orion in relation to the vertical from both ends of the chain simultaneously using a zenith sector. By their estimate, after building two observatories at either end and taking the necessary readings, they would be finished in another six months. This duplication of efforts necessarily led to further delays, as it required two teams of astronomers, the construction of a second zenith sector (in addition to the one they had brought from France), synchronisation of observations over 300 kilometres apart and, not insignificantly, clear night skies at both locations. Astonishingly, it would be another three years before the measurements were finalised and it would not be until 1744 before the mission finally returned to Europe.

The difference between the two observatories at the northern and southern ends of the chain amounted to 3°07'01" of latitude. The distances having been determined by triangulation, the scientists calculated that one degree of latitude on the equator amounted to 110.6 kilometres. The other expedition to Lapland found the difference to be 111.9 kilometres.

A survey between Paris and Amiens in 1669 had found that one degree of latitude in France measured 111.2 kilometres. Together, the three figures once and for all settled the argument between Newtonians and Cartesians. The Earth was oblate, that is, like an orange.

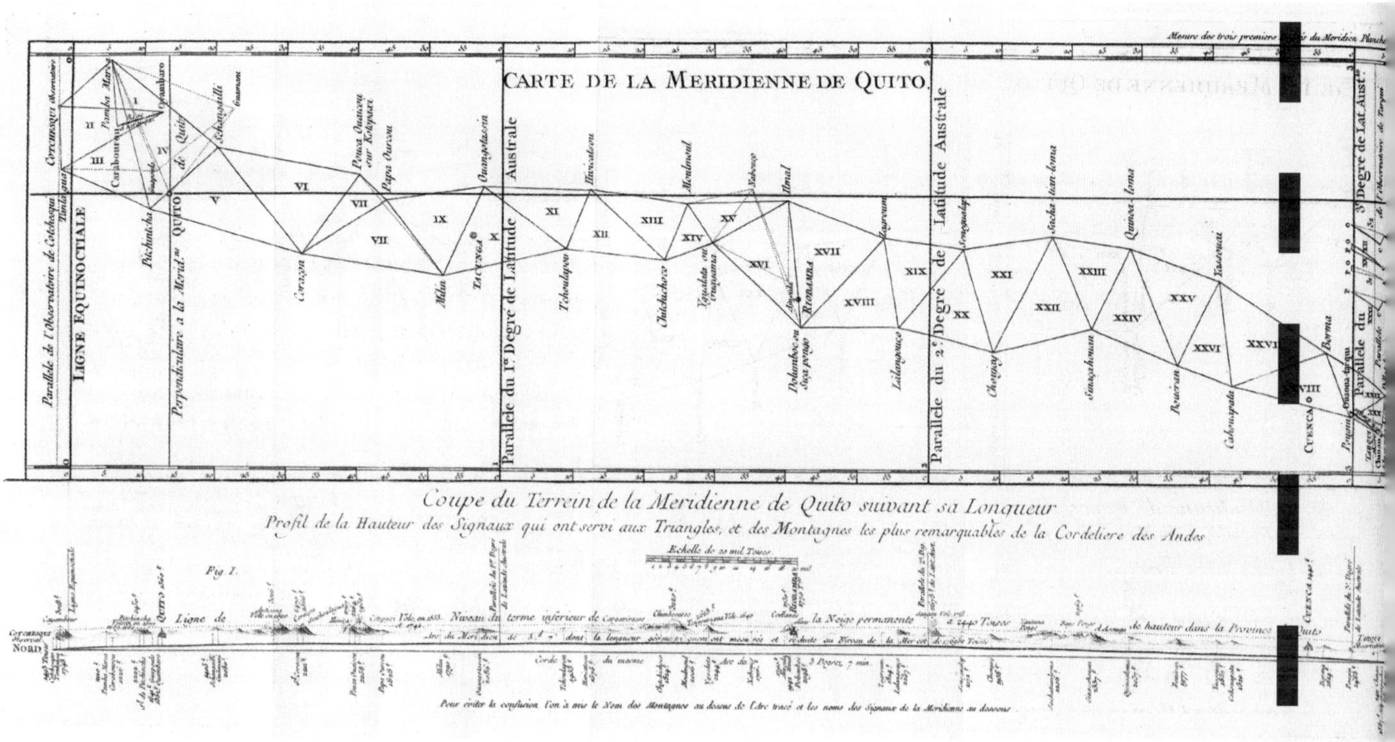

La Condamine's map and profile of the Quito meridian showing the height and position of the triangulation stations and the chief mountains along the Andean chain.

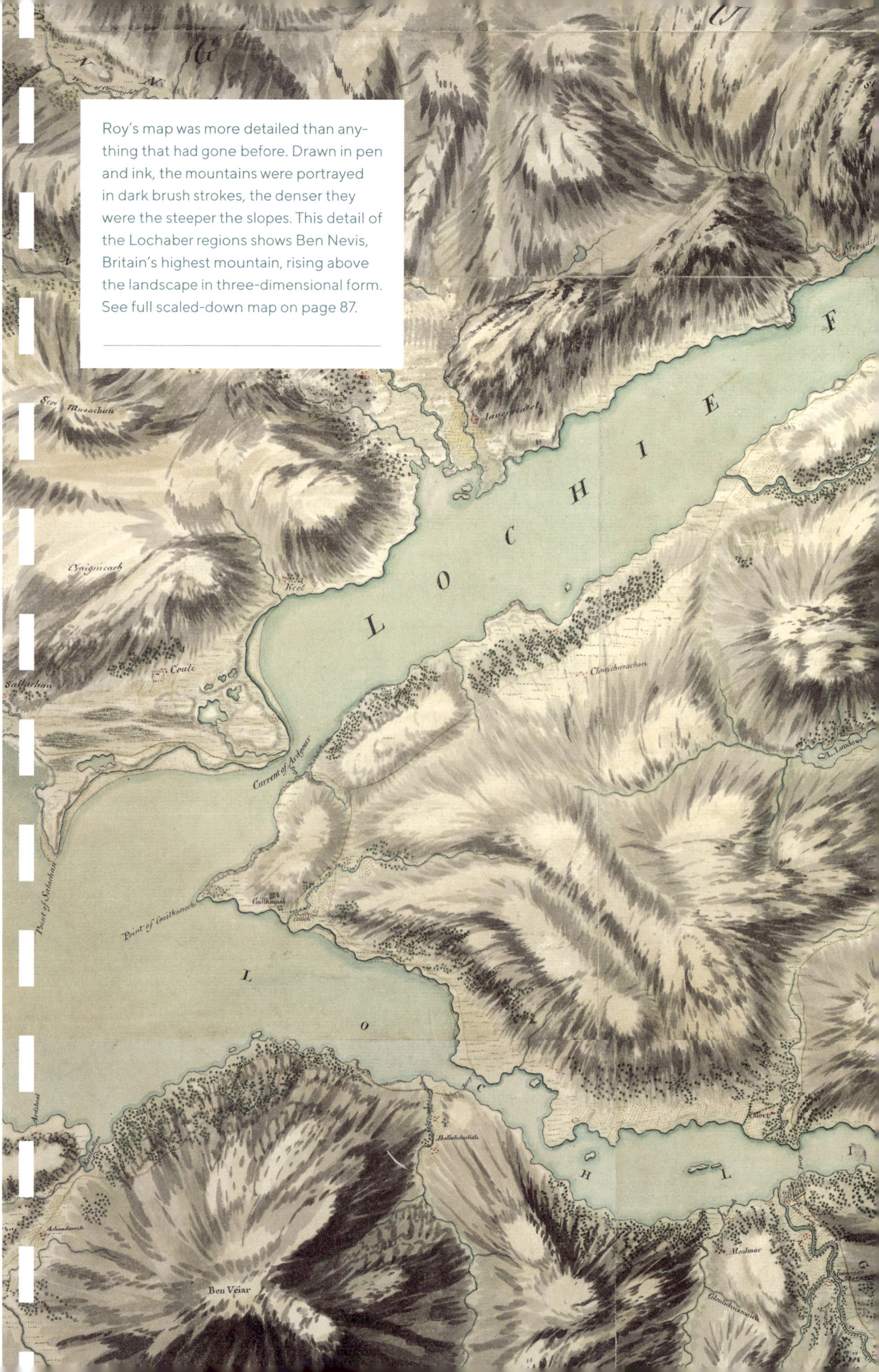

Roy's map was more detailed than anything that had gone before. Drawn in pen and ink, the mountains were portrayed in dark brush strokes, the denser they were the steeper the slopes. This detail of the Lochaber regions shows Ben Nevis, Britain's highest mountain, rising above the landscape in three-dimensional form. See full scaled-down map on page 87.

Lochaber
Fort William
Ben-Nevis
Skur Cherrin
Skurnacherie
Caloch
Gar-Ben

77

William Roy's Great Map

The origins of Britain's national mapping agency can be traced back to an ill-fated rebellion in the Highlands of Scotland

One summer's morning in 1747, a young surveyor acting in the king's service strode out from his barracks in the Great Glen with instructions to measure the roads around Fort Augustus. It was a precursor to a country-wide military survey. Having been plucked out of relative obscurity to work on the project, he must have had every reason to feel apprehensive, working with no on-the-ground military experience and in a hostile environment. Little more than one year before, on a bleak expanse of moorland less than 60 kilometres away, a bloody battle had been fought, its outcome triggering a reign of retribution that would irreversibly change the landscape of the Scottish Highlands which the young William Roy was now about to map.

The political and military turmoil in the Highlands could be traced back to the religious tensions at the top of the British establishment towards the end of the 17th century. Six decades before Roy first appeared in Fort Augustus, a Catholic monarch from the House of Stuart occupied the British throne. However, the reign of King James II of England (and VII of Scotland) who came to the throne in 1685 was not an auspicious one. It lasted three turbulent years before he was eventually deposed in 1688 and replaced by his Protestant daughter, Mary II. The unpopular James was banished to France, where the dethroned king maintained a court in exile, whilst furthering his so-called Jacobite[1] cause.

In the years that followed, Jacobite sympathies continued to simmer under the surface in Britain and occasionally boiled over into open rebellion. The two most significant uprisings, in 1715 and 1745, had their seats of sedition in Scotland and relied heavily on the support of the Highland clans in the mountainous north and west of that country, most of whom swore allegiance to the Catholic faith and the

Stuarts. Their relative isolation meant that the clans were a law unto themselves, lying *de facto* beyond the jurisdiction of either Edinburgh or London: the clan system was run on martial lines, where the landholding chieftains gave absolute protection in exchange for loyalty. If British law and order was ever going to be restored to this part of the realm and the threat of Jacobitism eliminated, it was imperative to root out this opposition, which – given the rugged terrain – was going to be a tough challenge.

In the second half of 1715, Jacobite forces had held much of Scotland in sway before being beaten back by troops loyal to the Protestant King George I, the first in a line of Hanoverian kings that ruled Britain following the demise of the Stuarts. In 1724, in the aftermath of the Jacobite defeat, General George Wade, who had helped quash the rebellion, was sent to Scotland to assess the extent of the surviving Jacobite menace in the Highlands. He reported back that the threat was real and recommended that the British government undertake a programme of road-building and fortification to counter this danger. Wade was immediately appointed *Commander in Chief of His Majesty's forces, castles, forts and barracks in North Britain* and formulated a plan to build hundreds of kilometres of roads, linking up strategically located garrisons and forts, including Fort Augustus – from where Roy set out that summer morning.

In many ways, the philosophy behind the strategy was not dissimilar to that of the Romans and nor was its implementation. The roads were to act as supply lines and for troop movements to deter a hostile population. Wade's roads were likewise built in straight lines, only deviating into zig-zags when the gradient so required. At regular 10-mile (or 16-kilometre) intervals there were military encampments and even inns. By 1745, Wade had been responsible for constructing about 40 bridges and over 400 kilometres of roads between the forts, three of which – Fort William, Fort Augustus and Fort George – linked up the southern and northern ends of the Great Glen, a deep geological gash that divides the Highlands in two. The roads may not have thwarted the 1745 uprising, but they were built to last and the network and its later expansion played a key part in the eventual subjugation of the clans. Indeed, today's Ordnance Survey maps still mark them as 'General Wade's Military Roads'.

In August 1745, Prince Charles Edward Stuart, grandson of James II, and popularly known as Bonnie Prince Charlie, sailed from France to raise the Jacobite standard in Glenfinnan in the Scottish Highlands. Amassing the support of Highland clans and other sympathisers along the way, by December the Prince's army had marched on Derby in the English Midlands, causing some degree of panic in London. However, by the turn of the year, British troops advancing northwards had gained the upper hand, pushing the rebels back into Scottish territory. It came to a final showdown between the two armies on 16 April 1746 at the Battle of Culloden, near Inverness. It was not so much a defeat as a rout for the insurrectionist cause. Not only did it signal an abrupt end to Jacobitism, it gave rise to a prolonged period of repression in the Scottish Highlands, the repercussions of which would be felt for over more than a century.

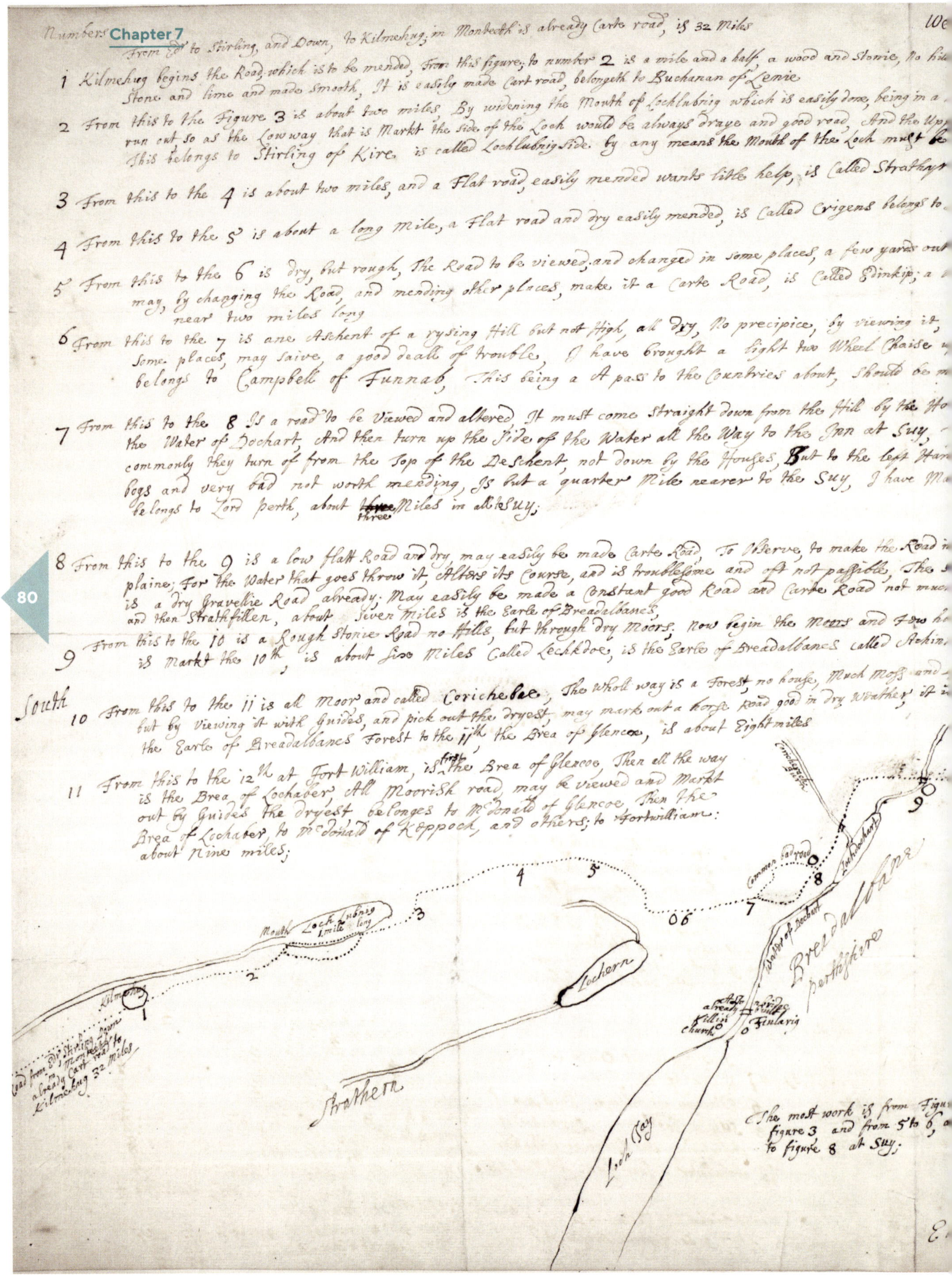

Numbers From Edr to Stirling, and Down, to Kilmehug; in Monteeth is already Carte road, is 32 Miles

1 Kilmehug begins the Road, which is to be mended, From this figure; to number 2 is a mile and a half a wood and Stonie, No hill
Stone and lime and made Smooth, It is easily made Cart road, belongeth to Buchanan of Lenie

2 From this to the Figure 3 is about two miles By widening the Mouth of Lochlubnig which is easily done, being in a
run out so as the Lowway that is Markt the side of the Loch would be always draye and good road, And the Up
this belongs to Stirling of Kire is called Lochlubnig Side. by any means the Mouth of the Loch must be

3 From this to the 4 is about two miles, and a Flat road, easily mended wants little help, is Called Strathyr

4 From this to the 5 is about a long Mile, a Flat road and dry easily mended, is Called Crigens belongs to

5 From this to the 6 is dry, but rough, The Road to be viewed, and changed in some places, a few yards out
may, by changing the Road, and mending other places, make it a Carte Road, is Called Edinkip; a
near two miles long

6 From this to the 7 is ane Ascent of a rysing Hill but not High, all dry, No precipice, by viewing it,
Some places, may saive a good deall of trouble, I have brought a light two Wheel Chaise
belongs to Campbell of Funnab, This being a A pass to the Countries about, Should be m

7 From this to the 8 Is a road to be Viewed and altered It must come straight down from the Hill by the Ho
the Water of Dochart, And then turn up the Side of the Water all the Way to the Inn at Suy,
commonly they turn of from the Top of the Deschent, not down by the Houses, But to the left Han
bogs and very bad not worth mending, Is but a quarter Mile nearer to the Suy, I have Ma
belongs to Lord perth, about ~~three~~ three Miles in all to Suy;

8 From this to the 9 is a low flatt Road and dry, may easily be made Carte Road, To Observe, to make the Road i
plaine; For the Water that goes throw it, Alters its Course, and is troublesome and oft not passible, The
is a dry Gravellie Road already. May easily be made a Constant good Road and Carte Road not muc
and then Strathfillen, about Siven Miles is the Earle of Breadalbanes,

9 From this to the 10 is a Rough Stonie Road no Hills, but through dry Moors, Now begin the Moors and Few h
is Markt the 10th, is about Six Miles Called Lechkdoe, is the Earle of Breadalbanes called Achin

South
10 From this to the 11 is all Moor and called Corichebae, The Whole way is a Forest, no house, Much Moss and
but by Viewing it with Guides, and pick out the dryest, may mark out a horse Road good in dry Weather, it
the Earle of Breadalbanes Forest to the 11th the Brea of Glencoe, is about Eight miles

11 From this to the 12th at Fort William, is first the Brea of Glencoe, Then all the way
is the Brea of Lochaber, All Moorish road, may be viewed and Markt
out by Guides the dryest belonges to Mcdonald of Glencoe Then the
Brea of Lochaber, to Mcdonald of Keppoch, and others; to Fortwilliam:
about Nine miles;

The most work is from Figure
figure 3 and from 5 to 6, a
to figure 8 at Suy;

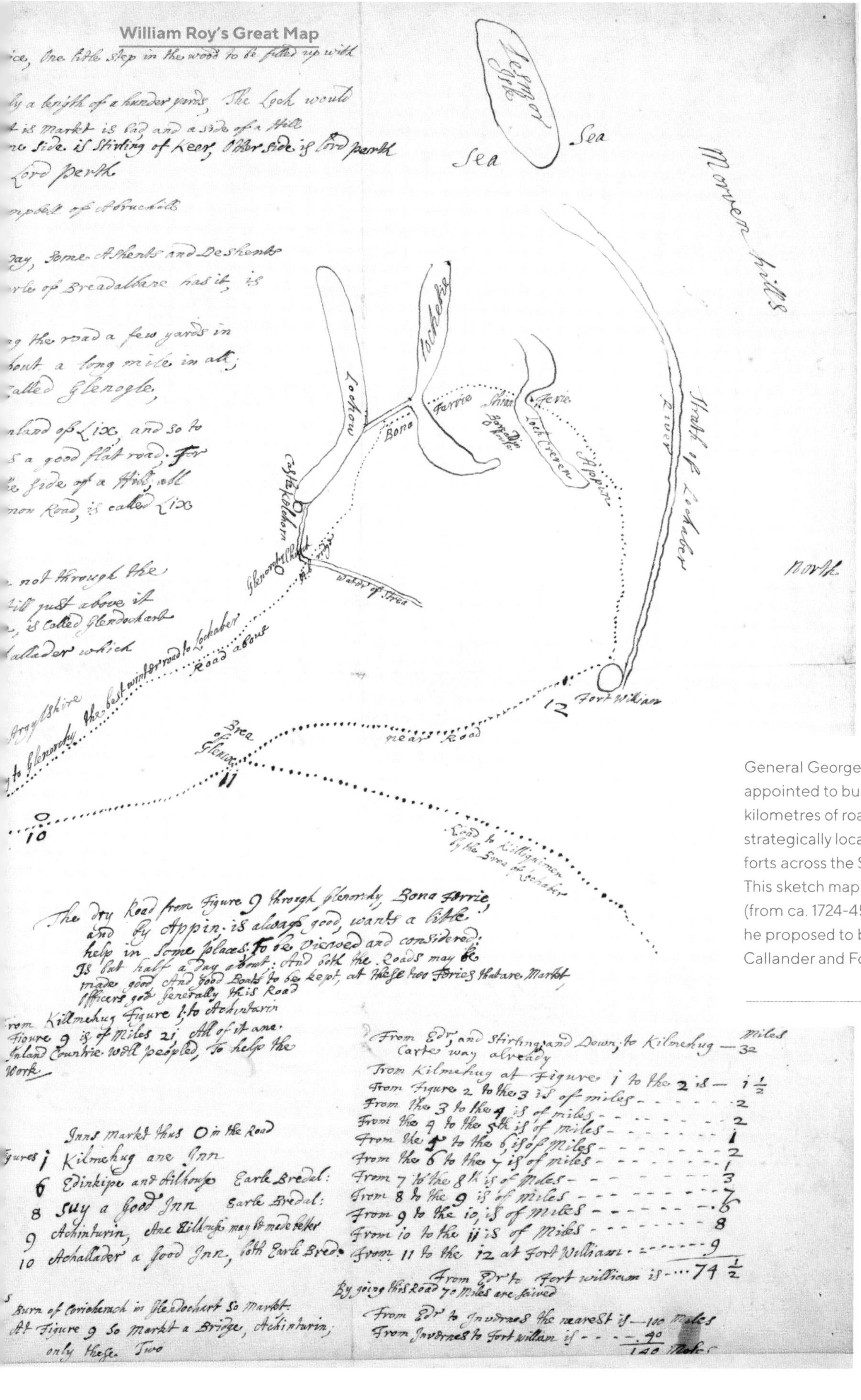

General George Wade was appointed to build hundreds of kilometres of roads, linking up strategically located garrisons and forts across the Scottish Highlands. This sketch map with annotations (from ca. 1724-45) shows the roads he proposed to build between Callander and Fort William.

The Battle of Culloden on 16 April 1746 ended the hopes of the insurrectionist Jacobite cause and set in train a regime of 'pacification' in the Scottish Highlands.

The response of the victors was one of retribution. The commander of the British forces that day was the 25-year-old Duke of Cumberland, the second son of King George II who had earned himself the name 'Butcher Cumberland' for his acts of brutality. From the very outset, the victory set in train a regime of 'pacification' in the Highlands which was resolved to destroying the clan system and wresting back control of the country.

To do this effectively, David Watson, the Deputy Quartermaster-General of the Board of Ordnance (amongst whose duties, in addition to maintaining and building harbours and fortifications, was mapmaking), was ordered to carry out an extensive military survey of the Highlands. The most up-to-date published map of Scotland available at the time of Culloden was a 1745 map drawn at a scale of 1:842,000 by John Elphinstone, a military engineer, who served with the Duke of Cumberland's army.[2]

Watson recruited William Roy to carry out an initial survey around Fort Augustus in 1747 to evaluate the feasibility of such a project. Roy was working for the Board of Ordnance at the time, and Watson knew him personally, as he had assisted in the survey of his family estate in the Scottish Lowlands. Despite his youth and inexperience, the 21-year-old, initially working on his own, proved an able servant. After a successful pilot project, Roy became the survey leader, and he was assigned several surveying teams, each consisting of an officer, a non-commissioned officer and six foot soldiers. The surveying season took place between the months of April and October during which period they would measure highways and byways, as well as the course of rivers, across the Highlands. Incredibly, between 1748 and 1751, after just four summer seasons, the whole of Scotland north of the Clyde-Forth line had been surveyed.

Roy used a method known as 'traverse surveying', which meant following a route – a road or the course of a river - using a pole, a chain and a surveyor's compass. A straight line along a road would be marked off with two poles – a back-station and a fore-station – the latter which would be positioned at the exact position where the route started to deviate (for example, a bend in the road). The distance between these would be calculated using the chain, and the angle at which the road deviated would be read using the compass fixed on a tripod. Before moving on, the compass would also be used to fix the angles of prominent features in the locality, whose relative proximity would be logged using educated guesswork. The surveyor would also draw an impression of the scene he saw in a field notebook.

Over the course of the day – and indeed for every day of the campaign – this would be repeated. Carried out over several seasons, it would have been a painstakingly long process. During the winter months, the field notes would, just as painstakingly, be interpreted by a team of draughtsmen working in Edinburgh and the data from six different surveying teams pieced together and arranged to slowly complete the jigsaw that was Scotland.

The result was a topographic map, drawn to a scale of one inch to 1,000 yards (1:36,000), greater in detail than anything that had gone before. It was drawn in pen and ink with a colour coding system – brown for roads, blue for water and green for woodland – which would later become familiar in Ordnance maps. The mountains were drawn in dark brush strokes, the denser they were the steeper the slopes. Many of the prominent peaks are named on the map. Because they were surveyed by sight from a lowly position, their forms are generalised, but some, such as Ben Nevis – at 1,345 metres, Britain's highest mountain - rise above the landscape in three-dimensional form.

By 1755, the whole of Scotland had been mapped in this fashion. The *Military Survey map of Scotland* 1747-1755, known by its contemporaries as the 'Great Map', went beyond Roy's original remit of the Highlands alone. It was unusual in respect of the fact that a national map had been realised, starting not in the metropolitan lowlands, but in the mountainous terrain of the Highlands. Eight years the wiser, Roy was aware of its shortcomings: it was no more than a military sketch, he was later reported to have said, useful only for the purposes for which it had been intended, not an accurate map. Traverse surveying had its merits, of course, and was a first attempt to systematise and standardise mapping, but Britain was still some way behind the French at a time when the Cassinis were gradually rolling out a more sophisticated series of maps based on triangulation and geodetic measurements. The military survey, for example, had made no astronomical observations to geo-reference locations by latitude and longitude. Nevertheless, despite all these shortcomings, a blueprint was now in place for the mapping of Great Britain in its entirety.

It's uncertain what direct influence Roy's military map had on the pacification of the clans after Culloden, nor the role it played in the inglorious Highland Clearances in the following decades. However, we do know that from a mapping point of view it was a landmark achievement, providing a unique picture of the Scottish Highland landscape prior to the industrial revolution. The map would also serve as a springboard for the future military mapping of Great Britain and Ireland. William Roy himself became a distinguished cartographer and in 1765 rose to the position of Surveyor General, the most senior post in this discipline at the end of the 18th century. He is generally recognised as being the father of the Ordnance Survey, Great Britain's mapping agency today. The military survey was just one of many important milestones in his illustrious career: later he would go on to establish the first baseline in Britain on Hounslow Heath near London and work in association with the Cassinis on a major triangulation exercise to calculate the relative longitudinal positions of the French and British observatories.

Endnotes

1 The name is derived from Jacobus, the Latin version of James. Jacobites were followers of the Jacobitism cause.

2 There were a number of more accurate military maps dating from the time of General Wade's roadbuilding programme, but these were local and piecemeal in nature and subject to official secrecy.

Paul Sandby (1731-1809) was a young artist who accompanied the surveying parties in the field, in part to document their operations. This sketch shows a survey party at work near Loch Rannoch in the southern Highlands in 1749.

The Elphinstone map of Scotland 1745.

By 1752, William Roy and his surveyors had produced finished maps of the whole of the Scottish Highlands, comprising 40,000 square kilometres of mostly harsh terrain, at a scale of 1:36,000.
The full map can be viewed on the website of the National Library of Scotland.

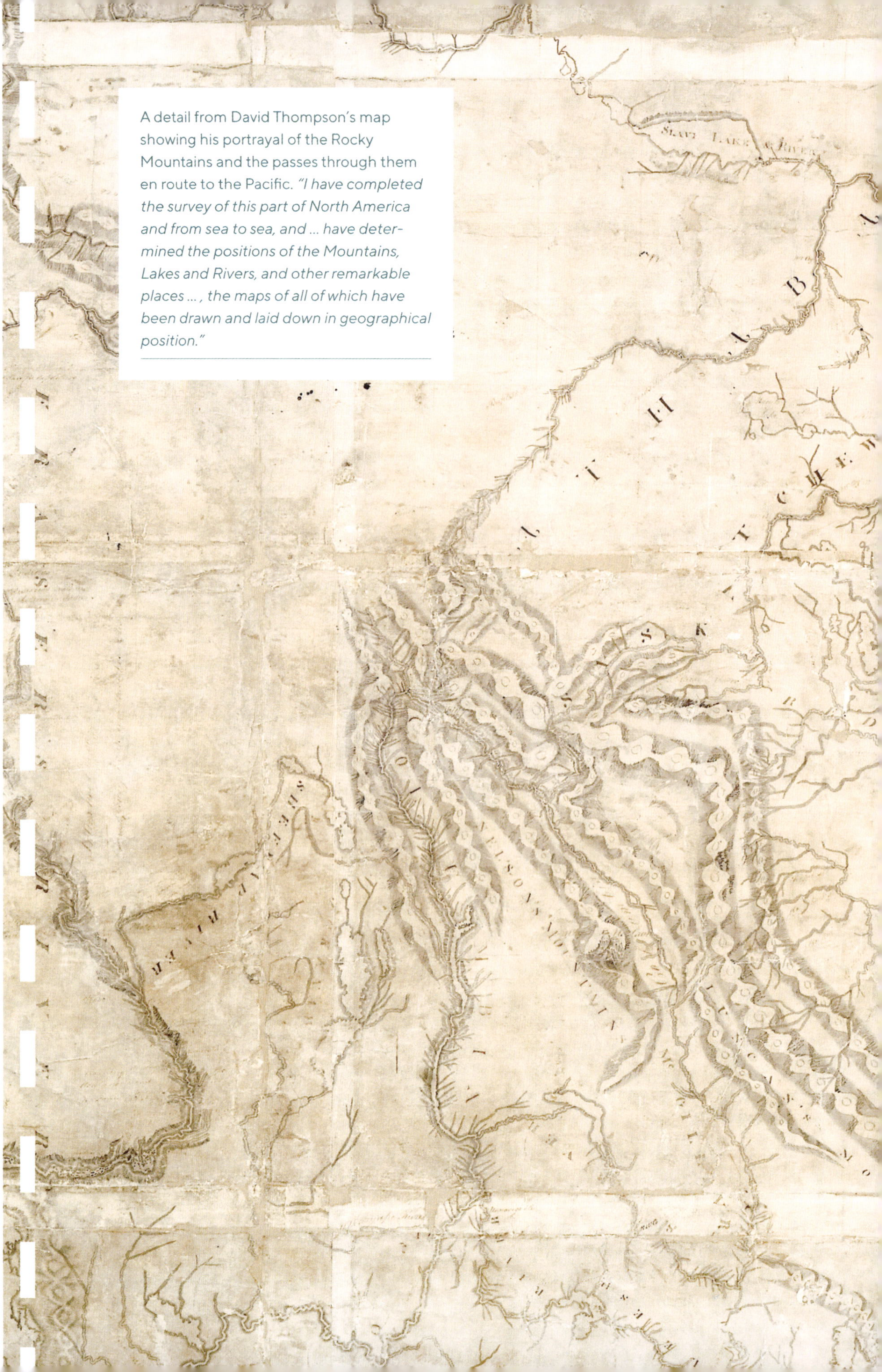

A detail from David Thompson's map showing his portrayal of the Rocky Mountains and the passes through them en route to the Pacific. *"I have completed the survey of this part of North America and from sea to sea, and ... have determined the positions of the Mountains, Lakes and Rivers, and other remarkable places ... , the maps of all of which have been drawn and laid down in geographical position."*

The stargazer

David Thompson was almost single-handedly responsible for mapping the westward advance of European trade and exploration across the Canadian Rockies

I am getting tired of such constant hard journeys; for the last 20 months I have spent only barely two months under the shelter of a hut, all the rest has been in my tent and there is little likelihood the next 12 months will much be otherwise.

From the comfort of our centrally-heated world, it is difficult to imagine the privations of a fur trader in the Canadian backwoods at the turn of the 19th century. Sleeping outdoors was but one of these. Voyageurs who plied the interior in their canoes, had to put up with searingly cold winters and endure the scourge of the mosquito in the summer. Going hungry for days on end for lack of food was not uncommon and if they were ensnared by the weather before reaching their wintering quarters, erecting a log cabin and holing up for several months hundreds of miles from home was the only means of survival. And if these hardships didn't defeat them, their own demons might.

Our story starts with the wide-brimmed felt hat whose popularity in late 16th-century Europe fuelled a demand for beaver pelts which could be supplied in abundance from the North American colonies. Over the previous century, the French had forged trading ties with the indigenous peoples of the Canadian interior, with furs changing hands for European wares, such as textiles and metal goods. By the end of the 17th century, fur-trading operations were centred on the hub of Montreal on the Saint Lawrence River and over time a whole network of trading posts developed inland along the rivers of the continental heartland to streamline the seemingly endless supply of profitable furs. In competition, the British founded forts along Hudson Bay to open up routes into the more northerly interior.

The drainage system across the Canadian Shield, carved out by millions of years of glaciation, was highly conducive to waterborne traffic, and the Great Lakes and the rivers, such as the Saint Lawrence, Nelson and Saskatchewan, became the highways along which the goods were moved. Non-passable stretches of water, such as steeply inclined rapids and watersheds between river systems, were connected by overland 'portages'.[1] The vessel of choice for transporting crew and goods was the traditional canoe, made from skinned birch bark and a light wooden frame. These were light enough to be portaged, yet sturdy enough to carry several tons of cargo, even in shallow water. Although susceptible to damage from rocks, they were easily repaired and their versatility ensured that they quickly became the favoured method of transport by European voyageurs.[2]

Towards the end of the 18th century two companies dominated the fur trade – the Hudson's Bay Company (HBC) and the North West Company (NWC) – and both were keen to expand their operational territories westward into unexplored regions inland, brought on primarily by the depletion of furs in the east of the country. To improve efficiency and profitability this required knowledge of the geography and the river systems. What was needed were individuals with a pioneering spirit and the navigational skills to do this. In 1778, the HBC were the first to appoint an inland surveyor: Philp Turnor. He was tasked with visiting newly established trading posts inland from the company's York Factory base, at the foot of the Nelson River on Hudson Bay. Amongst other things, this involved calculating longitude and latitude and determining distances between the settlements.

Six years later, David Thompson joined the company. He had been born in London 14 years previously to Welsh immigrants. After his father died when he was two years old, his widowed mother struggled with his upbringing and aged seven, he was placed with Grey Coat Hospital, a charity school in London. It was from here that he was recruited as an apprentice by the HBC and boarded the *Prince Rupert*, a ship bound for Canada, swapping the hustle and bustle of England's capital for the barren wastes of Churchill, a small, isolated fort and trading post on Hudson Bay. The contrast could not have been starker. "*I bid a long and sad farewell to my country, an exile for ever*," he later wrote.

In charge of the fort at Churchill was Samuel Hearne, who at one point during Thompson's first winter instructed the novice to transcribe some pages of a manuscript he was working on: *A Journey to the Northern Ocean*. These were the records of an epic 19-month overland trek Hearne had made in 1771, from Churchill to the shores of the Arctic Ocean, a round trip on foot that totalled 5,600 kilometres. His tales of endurance must have left a considerable impression on the teenager.

During his first years at Churchill, Thompson experienced at first hand the harsh winters and mosquito-ridden summers. He made several expeditions into the interior and spent the winter of 1787-88 with a team of fur-traders at an encampment of the Peigan, a nomadic tribe who controlled an area east of the Rockies in present-day Alberta. It was here that he was able to immerse himself in the language and culture of the First Nations.

David Thompson taking an observation.

Later that year, Thompson suffered a calamitous injury that would have a profound bearing on his future, falling down a steep river bank and breaking his leg (He later wrote that it was the best thing that ever happened to him.) His convalescence lasted more than a year and during the winter of 1789-90, encamped at Cumberland House, he made acquaintance with Philip Turnor, the HBC's chief surveyor. They soon struck up a friendship and Turnor introduced Thompson to practical astronomy, a subject in which the latter developed an avid interest. On the long dark evenings, he would borrow Turnor's instruments and take readings from the stars to establish precise latitude and longitude positions of Cumberland House.

This fortuitous encounter (and an unsolicited reference written by Turnor to the HBC) resulted in Thompson – at the end of his apprenticeship – receiving *"a brass sextant, a pair of parallel glasses ... and nautical almanacs"* from England in 1791. From now on, wherever he went, his surveying equipment would accompany him as he criss-crossed the vast hinterland that lay beyond Hudson Bay.

He was fastidious in his measurements and took astronomical readings at every stopping place along the way to identify its exact geographic coordinates. In between these stopovers, whenever feasible along the more peaceful waters, he would record – by means of traverse survey – every change in direction and make best-guess estimates of distances covered. On clear nights, he would assiduously observe the movements of the moon and the stars, using the sextant and consulting the nautical almanacs: it was said he could read the stars better than most folk could read a book and the Kootenay people who populated part of the Rockies reportedly invested him with the name of *Koo-Koo-Stint*, or the Stargazer.

It is calculated that by 1797, with the waterways as his guide, Thompson had covered a distance of almost 19,000 kilometres in the service of the HBC. During his travels, he had surveyed huge swaths of boreal wilderness, stretching from Hudson Bay to Lake Winnipeg and Lake Athabasca, and along the Saskatchewan River. Thompson had been offered promotion, but fearing that this would prevent him from further exploration, he took another momentous decision and defected to the rival North West Company.

The NWC operated out of Montreal and had established a significant network of trading posts from the Great Lakes to the Rockies, penetrating into the vast Athabascan region in the North West. Under the terms of the 1797 Jay Treaty signed by Britain and the newly formed United States of America, it was generally accepted that the border between the two countries would run along the 49th parallel, at least as far west as the Rockies. Part of Thompson's immediate remit was to carry out a survey to establish what part of NWC's operations lay south of this parallel and to explore the major rivers and lakes which straddled this arbitrary line, including the Missouri river, the headwaters of the Mississippi and the shores of Lake Superior. By May of 1798, Thompson had carried out the first mapping of the future states of North Dakota and Minnesota.[3]

Over the following decade, the more pressing issue for the NWC was to find a practical route across the nigh impassable Rocky Mountains and open up the region to the west of the mountain range as far as the Pacific, not only to the fur trade, but to ward off American interests. Thompson had been the man selected by the NWC to mastermind a project to seek out and navigate the course of a great river which flowed westwards and discharged into the Pacific Ocean. Its possible outlet had first been identified by an American sea captain Robert Gray, who named it after his vessel, the *Columbia Rediviva*, which was the first to enter the mouth of the river.

In the early years of the new century, Thompson made several unsuccessful efforts to cross the Rockies. In addition to the physically demanding task of breaching the continental divide, another obstacle he faced were the Peigan, the First Nation's people who settled the lands east of the continental divide, and who were opposed to any plans by European traders to exchange firearms with the Kootenai, their sworn enemies on the other side the Rockies.

It was in the summer of 1807 – the Peigan having been sidetracked by a dispute with other warring neighbours – that Thompson finally made his way up the North Saskatchewan river into the heart of the mountain chain. He followed a trail which had been used by the Kootenai for trading across what is known today as Howse Pass (1,539 m).[4] After a difficult descent on the western side they eventually arrived at a large river flowing north: by his calculations, he was five degrees north and seven degrees east of the mouth of the Columbia, but at this point he had no way of knowing that this was in fact the very same river. Instead he chose to travel south and set up winter quarters in a wide valley close to the source of the Columbia. He spent the next years setting up trade links with the First Nations on the western side of the Rockies and surveying the tributaries of rivers feeding in to the Columbia.

Thompson was a prolific sketcher, but only ten of his panorama sketches have survived. The inscription "Saleesh Mountains South of the Lake" indicates that this was drawn in early 1812 in what is now present-day Montana.

By 1810, US fur traders were seeking to expand their business across the Rockies too and Thompson received instructions from the NWC to pre-empt the Americans and establish a trading post at the mouth of the Columbia River. It was far from plain sailing.

Feeling sore on account of the NWC's trading with their First Nations rivals across the mountains, the Peigan blockaded the approaches to Howse Pass, forcing Thompson to find a new, more northerly passage across the Rockies for trading. This time he traversed the Athabasca Pass (1,753m) at the headwaters of a tributary of the eponymous river, near present-day Jasper. It was 10 January 1811 and temperatures had plummeted to minus 30° Centigrade. Travelling on snowshoes and with dogsleds, his team suffered all the extreme hardships of a Canadian winter. Nevertheless, Thompson remained undaunted: when he reached the top of the pass, he recalled, "*It was ... a most exhilarating sight.*"

It took them seven days to reach the Big Bend in the Columbia River where they set up camp to see out the winter. Here, over the next weeks, they built a cedar-plank canoe which would eventually carry them down the Columbia to the Pacific Ocean where they arrived on 15 July, only to find that the Americans had built their trading post, Fort Astoria, north-west of modern-day Portland, Oregon.

Reaching the Pacific marked a culminating point in Thompson's peregrinations. After 27 years, he felt his work had been done:

I have fully completed the survey of this part of North America and from sea to sea, and by almost innumerable astronomical observations have determined the positions of the Mountains, Lakes and Rivers, and other remarkable places on the northern part of this Continent, the maps of all of which have been drawn and laid down in geographical position.

During the long winters, Thompson had worked on his maps, based on his meticulously tabulated records of astronomical and geographical observations he made from thousands of locations across the land mass. From 1812 to 1814 he created his magnum opus, the great map, which he titled, "Map of the North-West Territory of the Province of Canada From actual survey during the years 1792 - 1812." The map found pride of place in the great hall of NWC's inland headquarters at Fort William (Thunder Bay), and when the NWC merged with the HBC in 1821, it was brought to England where it was used to revise maps of Canada over the following decades. Thompson himself died in relative obscurity in 1857, whence his map and his achievements suffered a similar fate. It was not until his journals were published in 1916 that his star was reborn.

Over the course of his career with the HBC and NWC, Thompson travelled 90,000 kilometres, by canoe, on horseback, on snow shoes and by dogsled, using thousands of ancient trails and waterways marked out by the indigenous peoples. It is estimated he covered a fifth of the North American continent, which amounted to almost four million square kilometres of territory. In his lifetime, he witnessed the shifting balances of power across the west and the progressive demise of indigenous cultures. As a proponent of the fur trade and explorer, he was an agent of this change, however sensitive he was to these societies. Nevertheless, for better or for worse, David Thompson is now quite justifiably recognised as one of the greatest geographers of all time.

David Thompson's 'Map of the North-west Territory of the Province of Canada' found pride of place in the great hall of the North West Company's inland headquarters at Fort William (Thunder Bay). In 1821, it was taken to England where it was used to revise maps of Canada over the following decades.

Endnotes

1. Astonishingly, it is claimed that the whole of Canada – from east to west – can be crossed by water, the longest portage being no more than 20 kilometres.
2. Canoe-building was just one of the skills that was needed to survive the rigours of the interior: navigation, hunting, foraging, fire-making and shelter-building were amongst the essential bushcrafting skills that fur-traders acquired from the indigenous peoples who had subsisted on this continent for thousands of years.
3. Copies of Thompson's maps were later used by Lewis and Clark's crossing of the continent between 1804 and 1806.
4. For a surprisingly low crossing along the Great Divide, Howse Pass was later shunned by the fur traders and by the Canadian Pacific Railway in favour of the Athabasca Pass and the Kicking Horse Pass respectively.

Glacier du Mont de Lans
Glacier du Mont de Lans
Jodri
Glacier du Col
Glacier de la Selle
Tête du Replat
Tête de la Gandolière
Aiguille du Sorellier
Plaret
Glacier du Plaret
Aiguille
S.t Christophe
Tête du Graou
Tête de Mansaton
Tête du Rouget
Roche Blanche
Tête de l'Or
Tête de la Maye
les Étages
l'Aiguille
Montagne de l'Ours
Tête du Poriet
la Gr.de Aiguille
Tête de la Cabourne
Tête du Creuzet
Rocher de Lamaria
Glacier des Arias
Aiguille du Canard
Glacier des Fétoules
Tête des Fétoules
Glacier du Vallon
Aiguille des Arias
Aiguille d'entre Pierroux
Glacier d'entre Pierroux
Tête de l'Etre
Glacier du Chardon
Glacier de l'Etre
Glacier des Sellettes
Glacier du Fond
Olan Pic
Cime du Vallon
Gioberney

The French *Carte d'État Major* (1820-1866) superseded the outdated Cassini maps of the previous century and mapped France at a 1:80,000 scale. One of its specific remits was to map more faithfully the hitherto poorly charted mountain regions - in particular the Pyrenees and the Alps. Definitive maps of the peripheral regions of France, such as this one showing the highest peaks in the Massif des Écrins, only started appearing in the 1860s.

The unlikely mountaineer

Astonishingly, between 1828 and 1830, no fewer than five 3,000-metre peaks in the French Alps were scaled by military surveyor Capitain Armand Durand

There are few ranges that encapsulate mountain history, and in particular mountain mapping, quite so grandly as the Massif des Écrins. Tucked away in the Haut-Dauphiné region in the south-east corner of France, the massif is a jumbled, but relatively self-contained mass of peaks rising to 4,105 metres. Its seclusion, away from major centres of population, meant that it was one of the last regions of France to be systematically surveyed.

Writing in the 1860s, Edward Whymper (1840-1911) was entranced by the Écrins, describing its cliffs, torrents and gorges as unsurpassed and its deep and savage valleys as having no equal. "*The valleys,*" he wrote, "*are for the most part short and erratic [and] wander hither and thither, taking one direction for a few miles, then doubling back, and then perhaps resuming their original course.*" Distant perspectives, he observed, were rarely to be seen and without an accurate map it was difficult to find one's way around in the massif. "*The district is still very imperfectly known; there are probably many valleys, and there are certainly many summits which have never been trodden by the feet of tourists or travellers ...*"

Before Whymper arrived in the Écrins on a peak-bagging mission in 1861, the high peaks of the massif had barely been explored and mapped. His description reflects the imprecise understanding of the mountain range which had up to then largely eluded exploration by humans. At that point, high mountain mapping, as Whymper alludes, was an imperfect science.

The Peutinger map (see chapter 3) shows that the Romans, after crossing the Col de Montgenèvre, probably forged passages to the north and the south of the Écrins massif on their way to their possessions in Gaul. In the Middle Ages, the isolation

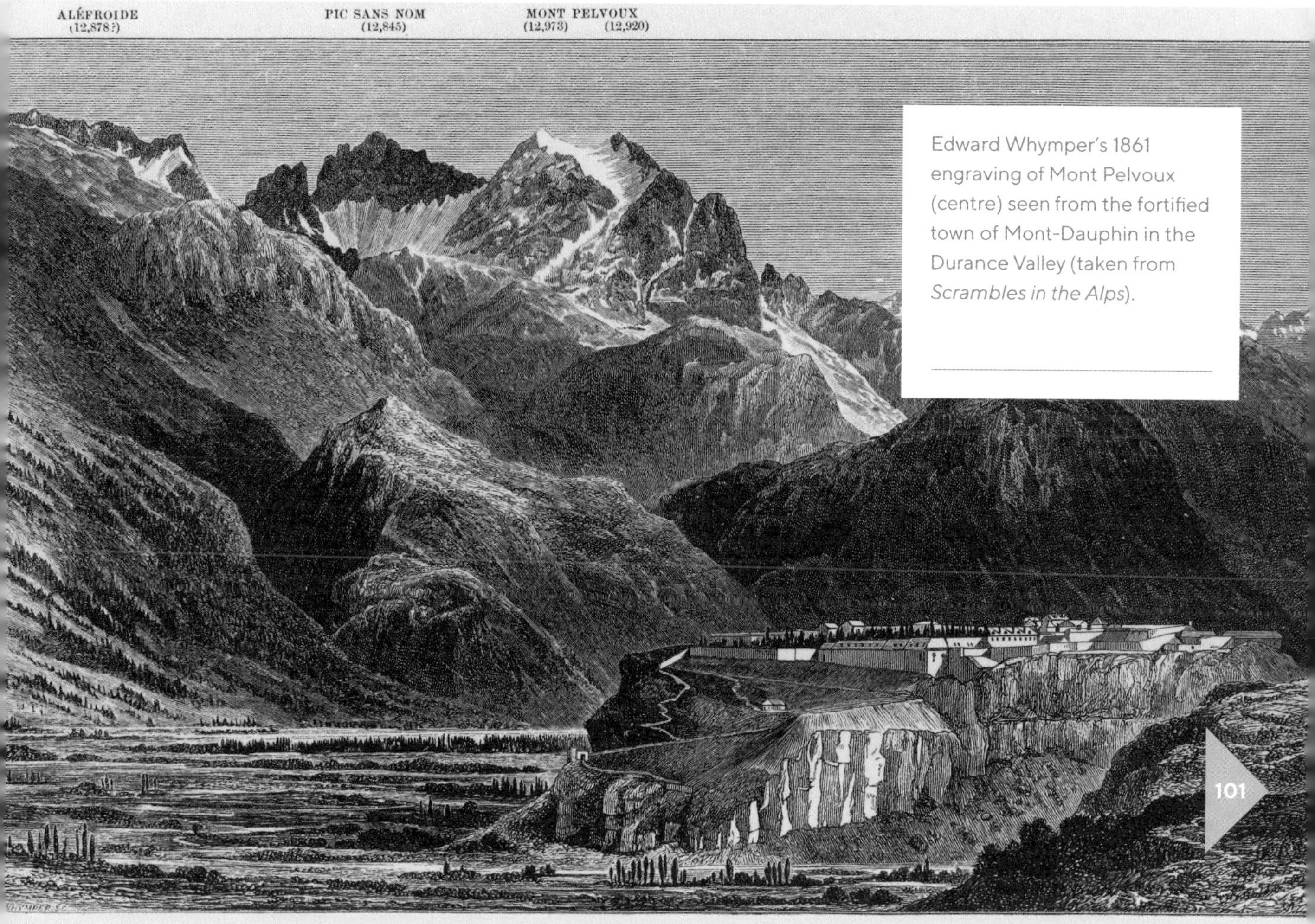

Edward Whymper's 1861 engraving of Mont Pelvoux (centre) seen from the fortified town of Mont-Dauphin in the Durance Valley (taken from *Scrambles in the Alps*).

of valleys in the massif and on the other side of the main Alpine watershed in Piemonte provided refuge to the persecuted groups such as the Waldensians, who emerged in the 12th century as a proto-Protestant community which fell foul of the Church of Rome.

Apart from the trading routes through the valleys and passes that skirted the Écrins, it wasn't until the early 17th century that anyone tried to make any sense of the region. Initial efforts to do so were in response to military rivalries between France and the neighbouring Duchy of Savoy.

The first detailed maps of the region were richly crafted manuscripts executed between 1604 and 1610 by Jean de Beins (1577-1651), a cartographer in the service of King Henry IV of France. For the time, this was cutting-edge cartography, but lacking the means and resources of later generations of surveyors, his work bears the telltale signs of early maps of mountain regions: the valleys along which roads pass are highlighted with the main settlements indicated according to their military and religious significance. The encroaching mountains are depicted in perspective with

The first detailed maps of the Les Écrins region were richly crafted manuscripts executed between 1604 and 1610 by Jean de Beins (1577-1651).

some relief shading and though there is some attempt to differentiate between the relative heights of the peaks, they remain indistinct and unlabelled, but are nevertheless enchantingly drawn.

It would be another century and a half before the cartography of the Écrins would be redrawn. By then, mapping techniques in France had advanced almost exponentially, inspired by a scientific impulse that had been sweeping the country, in particular under the reign of Louis XIV. In cartography, triangulation was enabling more geometrically exact maps produced under the direction of specifically trained *ingénieurs-géographes,* who could use increasingly lighter and more accurate surveying equipment.

The second part of the 18th century saw two important cartographic representations of the Écrins. In 1748, on behalf of the *Academie de Sciences,* Cassini de Thury (Cassini III) embarked on an ambitious public project to map the whole of France at a scale of 1:86,400, dividing the country into over 180 sheets. The project was the first of its kind to use the principles of geodetic triangulation. The first sheet (Paris, of course) was published in 1756. Sheet 151 (Briançon) – the one that covered the Écrins – was published in 1779. This survey was conducted by civilian *ingénieurs* who had to work in arduous circumstances to which they were unaccustomed, resulting on the one hand in a geometrically correct map, but on the other, in a topographically inaccurate one. Much of the high mountain relief on the Cassini map is sketchy, to say the least, as the surveyors preferred to observe from the periphery rather than from inside the massif.

The same cannot be said of the military mapping survey, which was carried out by Pierre Joseph Bourcet between 1749 and 1754. Bourcet was responsible for a team of 'rival' army *ingénieurs*. From all accounts the work was meticulously planned, with a baseline being set out in a flat section of the Durance valley near Briançon. From here they could then establish a network of landmarks, including mountain summits, for the purposes of triangulation. The surveyors climbed Vieux Chaillol (3,163m), a prominent peak in the south-west of the range, and crossed two passes of an almost equivalent elevation (La Muande and Le Says) in order to obtain their measurements. The 1:14,000 pen and ink manuscript maps that were the results of their endeavours were eventually reduced to a scale of 1:86,400. The maps predate the Cassini map by 20 years and are much more accurate and detailed, but because of their military sensitivity, they were not published until 1820.

The Bourcet map left out the highest peaks of the range which were of little use to the military, even less, in that respect, to mountaineers like Edward Whymper who came onto the scene a hundred years later. In his *Scrambles in the Alps*, Whymper writes: "*General Bourcet's [map] ... was completely wrong in its delineation of the mountains, and was frequently incorrect in regard to paths or roads.*" Later in the same chapter, in which he describes his own first ascent of Mont Pelvoux (3,946 metres) in 1862, Whymper makes reference to yet another *ingénieur-géographe*, who not only beat him to its summit by no fewer than 34 years, but who was also responsible for the next and improved series of government-sponsored maps of the region: Captain Adrien Armand Durand.

The *Carte d'État Major*[1] (1820-1866) was a nation-wide mapping project intended to succeed the Cassini maps and map France at a 1:80,000 scale. One of its specific remits was to map more faithfully the hitherto poorly charted mountain regions, in particular the Pyrenees and the Alps. The survey and mapping work would be done in stages, starting with a so-called first-order triangulation. This involved establishing a network of principal triangulation stations across the country to determine elevations and coordinates

Detail from Bourcet's 1:14,000 pen and ink manuscript map of the Les Écrins region.

Detail from the Briançon sheet of the Cassini map, published in 1779.

of these most prominent points. As with the earlier surveys, the high peaks of the Écrins would throw formidable hurdles in their path.

Once this gruelling phase had been concluded, an even denser mesh of second- and third-order triangles would be superimposed to home in on the detailed mountain topography so that the map could be completed. It was a long-term project.

The *ingénieur-géographe* chosen to commence the first-order triangulation in the Écrins in 1828 was the 41-year-old Captain Adrien Durand (1787-1835). After qualifying as a cartographic surveyor in 1810, Durand had followed the traditional career path within the *Dépôt de la Guerre*, the agency responsible for France's mapping. He showed great promise during the Napoleonic period, a prolific spell of mapmaking, having helped map the *Départements réunis* (in present-day Belgium) after they had been 'annexed' by the French Republic. From 1818 onwards he was employed on the *Carte de France* and earned renown for completing the Alpine part of a triangulation parallel that ran roughly from the Atlantic coast to the Italian frontier.

Between 1828 and 1830, Durand spent three summers in the Écrins taking measurements for the first-order triangulation. Accustomed to low-level surveying, the scale of the operations in the mountains was largely lost on the *Depot de la Guerre* back in Paris. In the Écrins, Durand was faced with huge obstacles: in order to set up first-order triangulation stations, which would have to be manned for days at a time to make accurate observations, teams of survey staff would be required to climb the highest peaks with cumbersome equipment, in what was unknown and perilous terrain subject to extreme changes in weather. The process was laborious and involved reconnoitring the peaks that would offer the best line of sight. Once these had been identified, a team of surveyors would have to navigate their way to the top and establish a triangulation station. A surveying party would then follow and build a signal and remain up there for several consecutive days to take measurements.

From a mountaineering perspective, Durand's overarching achievement during his work in the Écrins was to ascend Mont Pelvoux, at a height not thus far knowingly attained in France.[2] Although Durand must have been an experienced cragsman by the end of his tour, the prospect of ascending a mountain of this magnitude involving rock faces and ice fields would have been seriously daunting. For such tests of endurance, Captain Durand and his deputy Second Lieutenant Leclerc would have hired local 'guides' and for Mont Pelvoux this was no different. In order to tackle this peak, Durand engaged the services of two chamois hunters, Alexis Leotard and Jacques Étienne Mathéoud.

They set out from the Celse Nière valley south of the mountain on 30 July 1828 and had to negotiate steep rock and snow slopes before cautiously crossing the Clot de l'Homme glacier and then reaching the summit plateau which afforded access to two peaks of almost equal height. Durand must have come to the conclusion that the easternmost summit, a shade lower than its western twin, afforded the best sightings. It was here, at 3,932 metres, that Durand decided to install his signal. At this point we should remember that this was patently not an exercise in mountaineering, but part of a wholly more time-consuming undertaking, that of measuring the terrain. No sooner had Durand completed the first ascent of Mont Pelvoux than he sent for a team of surveyors and porters, and over the next few days they constructed a summit signal comprising a sturdily built cairn with an opening for a theodolite. He returned to the peak in August 1830 to take further measurements and slept in a stone shelter he had had built on the slopes beneath the summit, where he was holed up for three days on account of snow, with just a canvas roof for shelter.

It is estimated that Durand himself climbed forty peaks over 2,500 metres and five over 3,000 metres during his time in the Écrins. By the time he had completed the first-order triangulation, he was not just a surveyor, he had become an accomplished mountaineer too. Over his three summer campaigns, he probably spent more than two hundred nights close to the sixty or so signals he installed. The two highest peaks in the massif, Barre des Écrins and La Meije, he deemed to be inaccessible[3] and it was from Mont Pelvoux that Durand was able to establish that, at 4,105 metres, Barre des Écrins was the highest (at the time) in France.

Now well into his forties, the immense task of this first-order triangulation had taken its physical and mental toll on Durand. He also felt his achievements in the Écrins had largely gone unappreciated by his superiors in Paris. His requests for promotion to the rank of Commandant went unanswered. His disillusionment was exacerbated by headaches and fever which he had been suffering since 1831 and he was ultimately admitted to a mental asylum where he died in January 1835. It has recently been suggested that his symptoms were caused by the final stages of syphilis.

The 1:80,000 Briançon sheet of the *Carte d'État-Major* only appeared in 1866, 31 years after Durand's demise.[4] In the intervening period, second-order triangulation was carried out to create additional triangles within the first-order triangles, followed by an even more intricate third-order triangulation. Cartographic drawing – in the field – then took place in manuscript form at a scale of 1:40,000 from which the final engraving work would be completed at 1:80,000.The engraving itself required some deft handiwork: not only did they have to downscale from the manuscripts, the increased density of hatching and relief shading to depict the declivity of the mountain terrain – hardly necessary in lowland regions – also had consequences for the legibility of a monochrome map. From start to finish, the work for the Briançon sheet would span no fewer than 38 years.

In the years before the map was published, interest in the Écrins only grew. After 1850, the mountains attracted new generations of tourists and mountaineers. The need for trustworthy large-scale maps of the mountains became compelling, yet the only accurate maps before the 1866 *Carte d'État-Major* were those drawn up by Bourcet. Many visitors, like Whymper, who came using this 1820 map, redrew the topography based on their own observations and inch-by-inch knowledge of the terrain improved as they explored. Other more resourceful publicists with contacts in the *Dépôt de la Guerre* were granted access to the final manuscript maps of the *Carte d'État-Major* and were able to produce a more exact representation of the topography, before the official one was even produced.

For the growing sport of mountaineering, even when the *Carte d'État-Major* finally appeared and consigned Bourcet's map to the archives, it was already inadequate for high-altitude pursuits. In 1874, the *Club Alpin Français* stepped in and launched an initiative to produce a map of the massif on a larger scale, in colour and with an improved representation of relief. They were given access to the original manuscript maps and an eye-catching 1:40,000 map, for the first time including contours, was published to coincide with the inaugural edition of the club's journal, the *Annuaire du Club Alpin Français* in 1874.

From now on, as in other countries which had their own mountaineering communities, the *Club Alpin Français* would assume the mantle when it came to mapping the high peaks, perhaps best embodied by the work carried out by Paul Helbronner in the first three decades of the next century (see chapter 13).

Sadly, Durand's colossal work in the Alps went unrecognised in official circles during his lifetime, but his legacy as a surveyor and mountaineer was gradually discovered by the mountaineering fraternity who came across traces of his work at the top of peaks they had thought were previously unclimbed. The lower summit of Mont Pelvoux from which Durand had taken his readings was christened Point Durand in his honour. His praises were sung by such notables as Edward Whymper, Henri Béraldi and Paul Helbronner[5] and in 1929, one hundred years after Durand's first ascent of Mont Pelvoux, the *Club Alpin Français* erected a plaque at the village of Ailefroide, in the shadow of the fabled peak, to commemorate his herculean efforts.

The author standing next to Durand's plaque at Ailefroide.

Endnotes

1 Literally, 'staff map' referring to the officers of staff responsible for the surveys
2 Until 1860, the Mont Blanc Massif was part of the Duchy of Savoy. The first ascent of Mont Blanc was on 8 August 1786.
3 In fact, the first ascent of Barre des Écrins was made by a party of Englishmen including Edward Whymper and the guides Michel Croz and Christian Almer on 25 June 1864. The highest summit of La Meije, known as Grand Pic (3,984 m), was first climbed on 16 August 1877 by local guide Pierre Gaspard and his son (also called Pierre) and their client Emmanuel Boileau de Castelnau. It was the last major peak in the Alps to be climbed.
4 This was also five years after Edward Whymper's ascent
5 Wikipedia lists Henri Béraldi (1849-1931) as a French bibliophile, publisher and author of books on the Pyrenees and on French printmakers of the 19th century. Edward Whymper and Paul Helbronner are referred to elsewhere in this publication.

Club Alpin français
la Grave
les Vernois
Villard-d'Arène
le Dauphin
Roche Mantel
Pointe de Muretouse
Glacier de Mont-de-Lans
Pic de la Grave
Jandri
Tête-Mouthe
Diable
Tête du Toura
le Râteau
Glacier de la Meije
Bec de l'Homme
Glacier de l'Homme
Glacier des Etançons
Glacier du Clot des Cavales
Glacier de la Selle
Tête du Replat
Tête de la Gandolière
Glacier du Plaret
Lanchatra
le Puys
Aiguille de Lanchatra
Roche Rumée
St Christophe
les Granges
Champhorent
les Etages
la Bérarde
Glacier de la Bonne Pierre
Glacier du Vallon
Montagne des Ours
Glacier des Arias
Glacier de la Mariande
Glacier d'entre Pierroux
Gd Vallon
Glacier des Fétoules
Glacier du Vallon
Tête de Chéret
Glacier de l'Etre
Gd Glacier de la Pilatte
Glaciers des Sellettes
Glacier du Fond
Gioberney
Pic des Opillous
Gravé par Erhard 12 rue Duguay-Trouin Paris.
Esquisse de la mise en relief par Mr Prudent Capne du Génie

In 1874, the *Club Alpin Français* produced a 1:40,000 map of the massif in colour and with an improved representation of relief, based on the original manuscript maps of the *Carte d'État-Major*.

Hoensbroek
Overbroek
Lothbroek
Ter Linden
Hellenbroek Veld
Ten Eschen
Weustenraedt
Terveeren
Weustenraeder Veld
Putter Veld
Voerendael
Kortenbach
Hoenshuys
Kunraede
Wingaertshof
Retersbeck

The 1:25,000 Tranchot maps, like this one of Heerlen (NL), paint a picture of the social and economic conditions of the pre-industrialised Rhineland and parts of the Netherlands in perhaps a more expressive way than any history book.

Napoleon's triangles

A prominent hilltop overlooking Aachen marked the starting point for a map of Napoleon's newly acquired possessions

Anyone trying to win the affections of their sweetheart could do worse than to walk them up to the summit of the Lousberg in Aachen at sunset on a fine summer's evening. Admittedly, it may not have the same ring to it as the Spanish Steps in Rome or Paris' Montmartre, but the view from the top of this hill just outside the city centre is as enchanting as any. Indeed, the unbroken views offered from this lofty vantage point were not lost on early cartographers either, as the prominent obelisk on its summit bears witness.

The story of the obelisk dates back over 200 years, to a time when military aggression was good for mapmaking. After Napoleon Bonaparte invaded the Rhineland at the turn of the 19th century, he demanded that detailed maps be made of his newly acquired possessions. The man chosen for the job was Jean-Joseph Tranchot (1752-1815), a French army officer who had already carved out a distinguished career as a cartographer.

During this period, France had taken the lead in systematic mapping. This lead was built on over a century of dedicated service to geodesy and cartography instigated by the Cassini family, an accomplished dynasty of astronomers and cartographers who, over the course of four generations from 1671 to 1793, occupied the position of *Directeur général de l'Observatoire* in Paris (see chapter 5). Between 1750 and 1793, largely under the supervision of César-François Cassini (known as Cassini III)[1], the first nationwide map of France had been produced, in a total of 182 sheets at a scale of 1:86,400. It was achieved using the most advanced scientific principles of the day, namely astronomy and triangulation.

The Lousberg in Aachen at sunset.

It was against this background that Jean-Joseph Tranchot started his career in the latter part of the 18th century. His first surveying commission took him to Corsica where he was to help with the triangulation of the island of which France had taken control from the Republic of Genoa. After two decades serving on this project, Tranchot - who had singled himself out for the great accuracy of his work - received a commendation from the French Academy of Sciences. Later he worked as a chief surveyor in the Méchain/Delambre measurement of the meridian arc between Dunkirk and Barcelona (1792-1798), which had the aim of establishing the exact size of the earth on which the new metric system would ultimately be based.[2]

Having a career that had spanned different regimes, Tranchot was clearly a man who not only showed a greater allegiance to his occupation than to the political order of the day, he was also someone who was held in high esteem by his masters. And so it was on the orders of Napoleon that in 1801 he was tasked with mapping the Rhineland. Tranchot set up his surveying operations in Aachen, the prefecture of the newly formed *Département de la Roër*, one of four 'United Départements' on the left bank of the Rhine stretching from Kleve (Cleves) in the north to Saarbrücken in the south.[3]

Tranchot's first mission was to establish the exact geodetic coordinates of the highest point of the Lousberg, a prominent hilltop overlooking the town. From here the whole of the surrounding country could be triangulated and mapped. To do this he took a series of meticulous astronomical observations from the summit over a period of three and a half weeks in the summer of 1803. His aim was to connect the Lousberg (264 m) to existing triangulation networks in France and the Netherlands and so mark the starting point for the new Rhineland network.

From here, Tranchot created a system of 55 triangulation stations throughout the Rhineland. In the lower-lying areas, measurements would have been taken from prominent landmarks, most commonly church steeples. Those in Sittard in the Netherlands and Erkelenz in Germany, which were in direct sight of the Lousberg, were amongst the first to be triangulated. They measured a distance of 28.12 and 36.59 kilometres respectively. However, in other parts of the countryside, such as in the Eifel, where higher points were covered in dense forest, wooden beacons had to be constructed.

The French instrument of choice was the so-called Borda's repeating circle, a contemporary alternative to the theodolite. Two distant points were measured using two telescopes mounted on a shared axis and the angle between the two measured on a calibrated scale. The accuracy of the readings could be increased by taking multiple, i.e. 'repeated' measurements.

Once primary triangulation in the Rhineland had been completed, topographic surveys could commence to map out the countryside in greater detail. This was done using plane tables, another standard piece of surveying equipment consisting of a drawing board mounted on adjustable legs and used in the field for plotting measurements directly. The results of these measurements were recorded in 'field maps' (first at a scale of 1:10,000 and later at 1:20,000). Draughtsmen would depict variations in relief (i.e. slopes) by the use of hachuring, where steeper slopes were represented by thicker, shorter 'hatching', while gentler slopes were represented by strokes which were thinner, longer and further apart. A very gentle slope or a flat area, like the top of a hill, was usually left blank.

For the Rhineland maps, cartographers were instructed to draw these hachures using the relatively new 'oblique illumination' technique, which introduced the principle of hill shading to cartography. In this way the terrain could be depicted in a more realistic fashion by showing how the three-dimensional surface might be illuminated from a point light source, for example, in the top left corner of a map. However, the cartographers of the day were not always properly acquainted with the technique. Others who had to work in haste did not always heed the requirements, which led to a certain lack of uniformity between the Tranchot maps.

The maps also included a cadastral element, so, for example, land was designated by its use: 'B' (*Bois*) indicated woodland; 'T' (*Terres labourables*) arable land; 'P' (*Prés*) meadowland; and so on.

After Napoleon's exile to Elba in 1814, the French withdrew from the Rhineland. From 1816 to 1828, the topographic survey of the region continued under Prussian rule, supervised by General Karl von Müffling (1775-1851). The cartographic series dating from the whole of this period is therefore known as the Tranchot/von Müffling series.

In all, 264 maps – each covering an area of 10 x 10 kilometres - were produced at a scale of 1:20,000 (later reduced to 1:25.000).

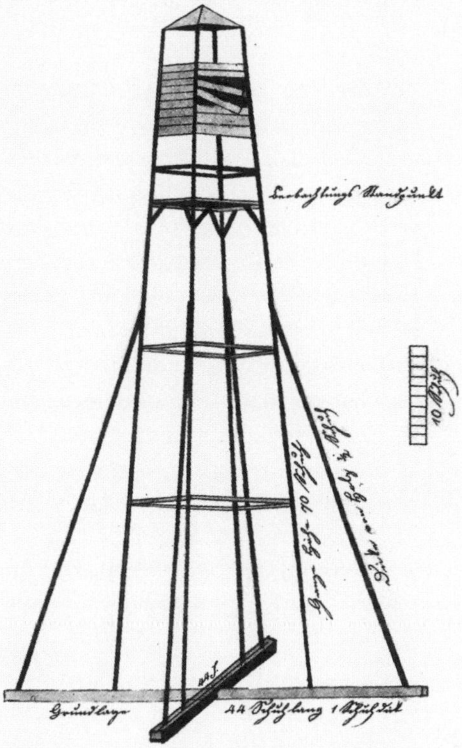

Drawing of a Tranchot beacon positioned at Altenburg (621m) in the Hunsrück.

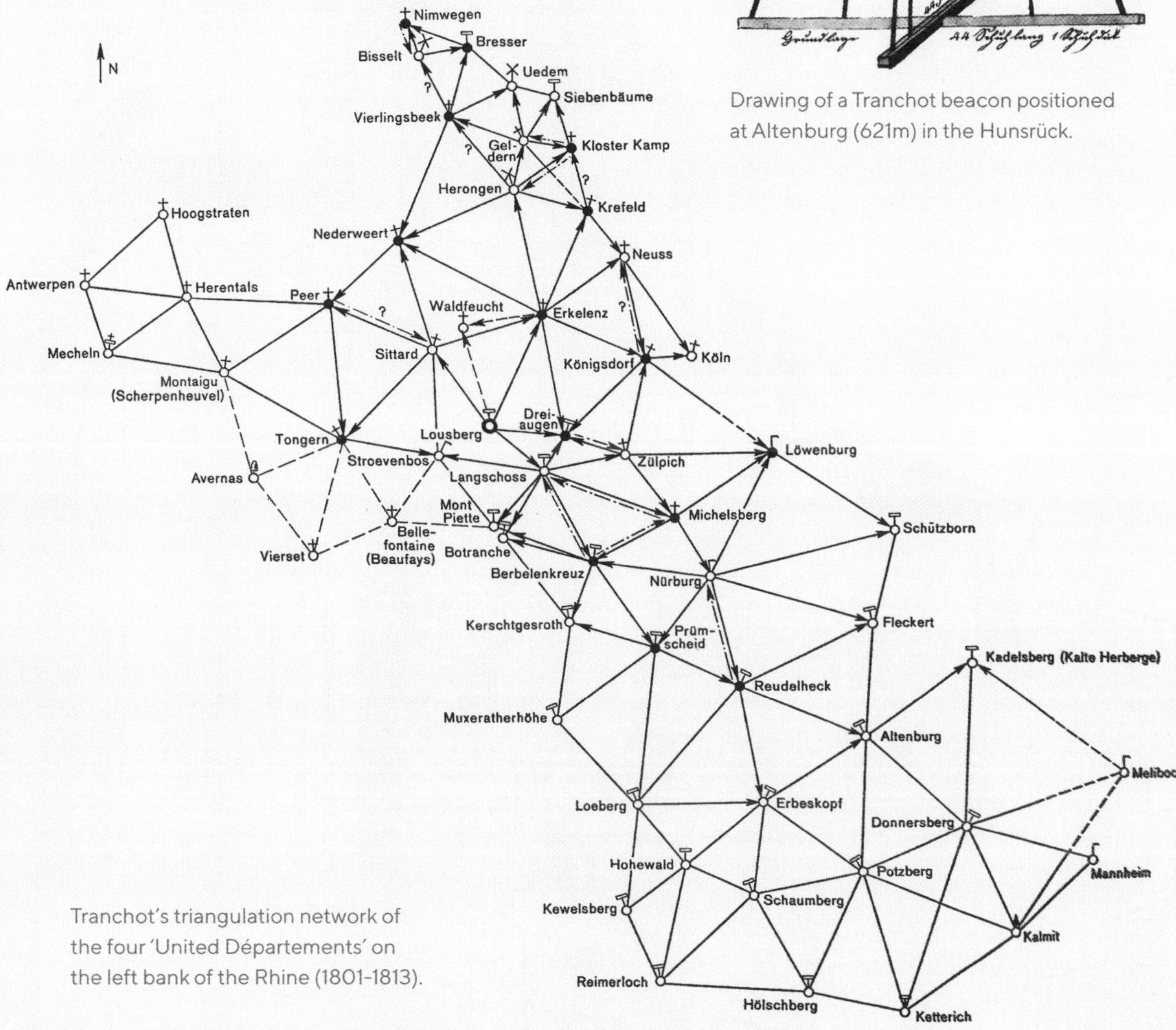

Tranchot's triangulation network of the four 'United Départements' on the left bank of the Rhine (1801–1813).

The Tranchot maps paint a picture of the social and economic conditions of the pre-industrialised era in perhaps a more expressive way than any history book. Agricultural reform was still in its infancy, individual settlements had not yet expanded and new roads between major towns and cities had not been built. Through a comparison with modern-day maps it's possible to understand the far-reaching changes that the landscape has undergone. On the other hand, in some remoter rural areas, it is surprising to see how much time has seemingly stood still for the last 200 years.

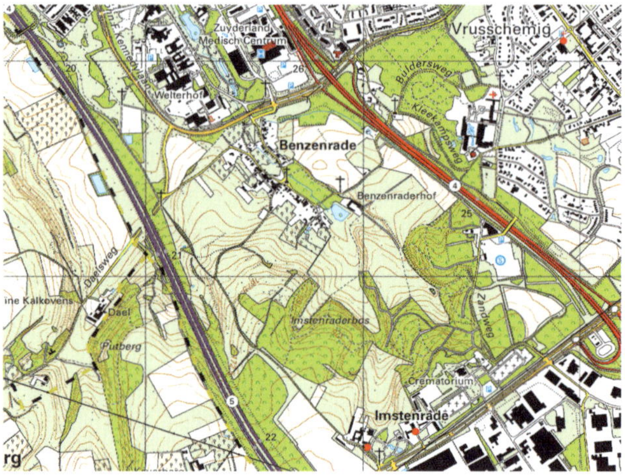

A comparison with modern-day maps can reveal the far-reaching changes that the landscape has undergone, but in some rural surroundings, it is surprising to see how much time has seemingly stood still for the last 200 years.

As for the Lousberg, the obelisk on the top of the hill tells its own story, one which captures the romance of cartography for map lovers. The triangulation point on the Lousberg was originally marked by a wooden signal. However, in 1807 a stone pyramid was erected on the spot and inscribed with various commemorative texts, not least proclaiming the accuracy of Tranchot's geodetic calculations. The square base stood 3.6 metres high and the pyramid which rose above it reached a height of 8.7 metres. Later that year a lead casket was placed in the base containing 55 gold and silver coins and 13 bronze medallions celebrating key moments in the life of the Emperor Napoleon.

On 2 April 1814, after French troops had withdrawn from the Rhineland, the obelisk was torn down by Prussian soldiers hell-bent on looting the valuables contained in the lead casket. In the process, several stones were damaged, some toppling down the steep slope at the top of the hill. In an act of generosity towards a fellow surveyor, Major-general von Müffling, the chief of general staff for the Prussian army in Niederrhein, immediately ordered the pyramid to be reconstructed, faithfully reproducing the texts that had been inscribed on the original pyramid, barring a eulogy in praise of the recently deposed Napoleon.

Happily, the monument still stands there today serving as a tangible record of the remarkable feats of cartographers of a bygone age. And if, while taking in the sweeping view from the top of the Lousberg, you were wanting to capture the attention of your heart-throb, what better way to do this than by telling them the captivating tale of Tranchot and his maps?

Military surveyors from the Grande Armée using a plane table.

Endnotes

1 César-François Cassini (de Thury) was the third generation of Cassinis who had occupied the position of director of the Paris observatory
2 In 1791, the French Academy of Sciences defined the new metre as being one ten-millionth of the distance from the equator to the North or South Pole: measuring part of a meridian arc (a line of longitude) between specific lines of latitude would define the size of the earth. Not surprisingly, France was the first to introduce the metric system (in 1795) and soon afterwards metrication was adopted for mapping purposes. Other European countries were quick to follow in the subsequent decades. Guillaume Henri Dufour published the first Swiss official map for which the metre was adopted as a unit of length, for example.
3 The offices he occupied have long since disappeared, but a plaque on the wall of the Kind-Jesu-Kapelle on the corner of Klappergasse and Jakobstrasse in Aachen commemorates Tranchot's stay in the city.

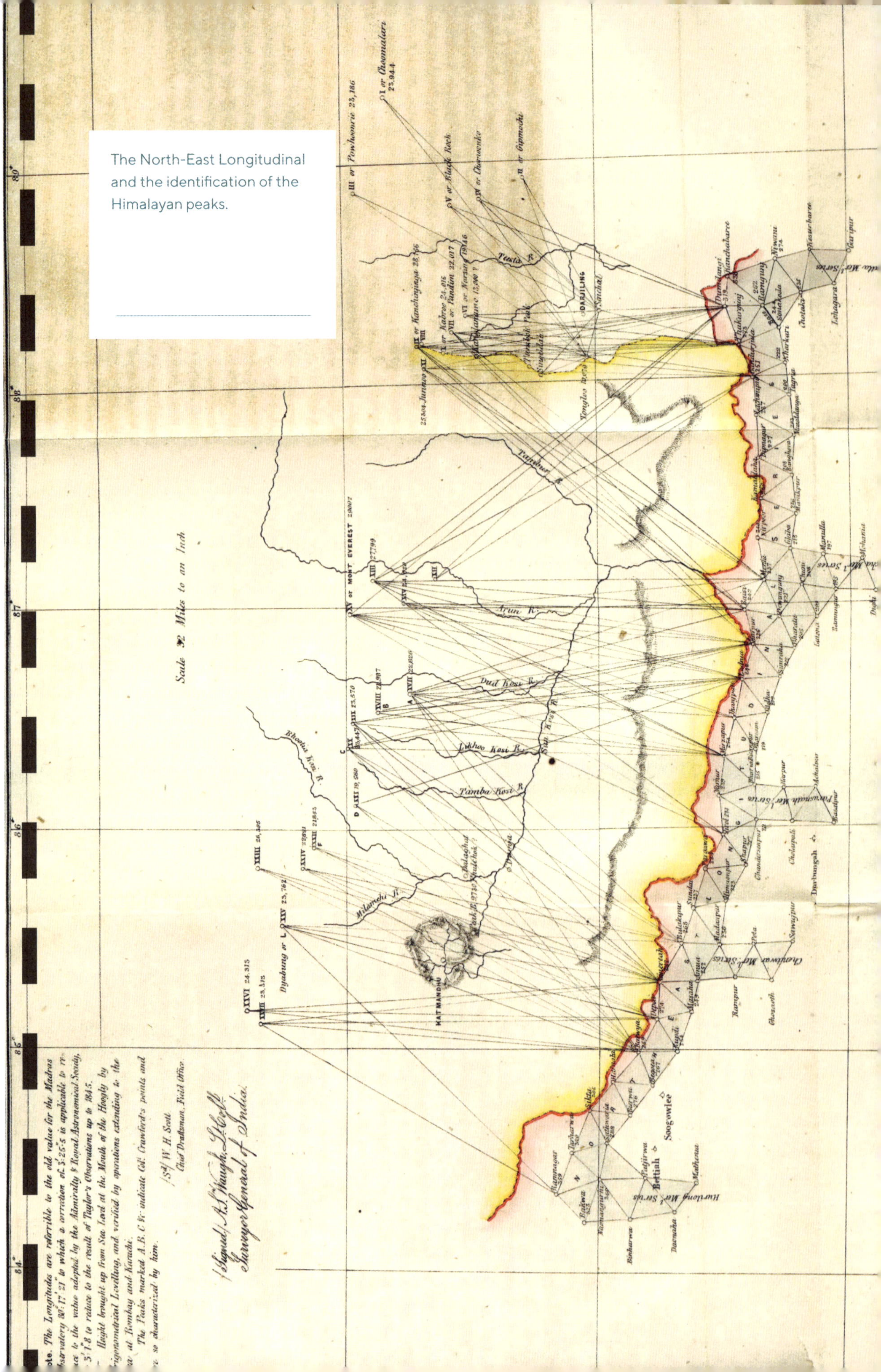

The North-East Longitudinal and the identification of the Himalayan peaks.

The Great Arc

Almost fifty years after the inception of the GTS, surveyors identified a peak on the distant horizon as the world's highest

The Great Trigonometrical Survey of India (GTS) was one of the greatest exercises in surveying and triangulation ever undertaken, ultimately culminating in the mapping of the Himalayas. It was instigated in 1800 by William Lambton, a British army officer and engineer serving in India who deemed that 'a mathematical and topographical survey of the greatest accuracy' of the Indian subcontinent was necessary. It was a recognition that, in respect of British-held territories on the Indian subcontinent, better mapping was required. Later, the project he began would also become an ambitious assignment in geodesy.

Lambton concluded an initial survey of the 13th parallel in the south of India in 1806. Proving his point, he established that the width of the Indian peninsula was shorter than on previously published maps. In its aftermath, Lambton embarked on a yet more challenging task which was to measure a 2,900-kilometre arc-of-the-meridian starting from the southernmost tip of India and working its way north to the Himalayas. It would become known as the Great Arc of the Meridian.

Responsible for initial funding, the East India Company thought the arc project would take about five years to complete, but eventually it took nearly 70 years to properly finish. Using a theodolite that weighed 50 kg, the survey work was toiling. Lambton worked with a team of 150 officers and assistants who had to sometimes hack through jungle and undergrowth on their way to the selected triangulation stations up and down the country. Team members were regularly plagued by fevers, injury and in some cases, death.

Despite these travails, such an extensive survey would add to the growing scientific body of information about the shape of the Earth, whilst also serving as a basis for mapping the subcontinent and ultimately establishing the Himalayas as the undisputed roof of the world.

Lambton died at the age of 70 in 1823 and by that time the survey of the meridian arc had only got as far as Nagpur, at the geographical centre of the subcontinent. His successor was his assistant George Everest, who proved of equal perseverance in fulfilling the task. Over a period of 20 years, Everest extended the survey of the Great Arc of the Meridian as far as the foothills of the Himalayas and it was there, in Dehra Dun, that he terminated its northward push.

In the meantime, measuring the Great Arc had itself become just one part of the Great Trigonometrical Survey of India. A surveyor's map of the sub-continent at the time resembled a spider's web of triangulation chains, known respectively as meridionals (running south to north) and longitudinals (east to west). The purpose of this 'grid-iron' was to stake out the rest of British-held India for mapping. Running parallel south of the Nepalese border with the Himalayas, for example, was the so-called North-East Longitudinal. This consisted of a 1,200-kilometre chain of triangles from Assam in the north-east corner of India to Dehra Dun. It was from the triangulation stations along this chain that the faraway snowy peaks to the north could – clear skies permitting – be observed.

In 1843, George Everest retired as Surveyor of India and was replaced by Andrew Waugh who was responsible for overseeing the GTS. Everest himself was known to be an admirer of such distant mountain panoramas, but was no peak-bagger, at least in surveying terms: to him, calibrating the heights of the Himalayan summits was almost a side-issue compared to the main task of mapping India. Waugh however took a keener interest, and in his plotting of the horizon by theodolite had identified Kangchenjunga as a prime candidate for the world's highest peak. Other contenders included Nanda Devi and Dhaulagiri.[1]

Recognising the fact that the Indian subcontinent required more accurate mapping, William Lambton carried out an initial survey of the 13th parallel between Chennai (Madras) and Mangalore in the south of India in 1806.

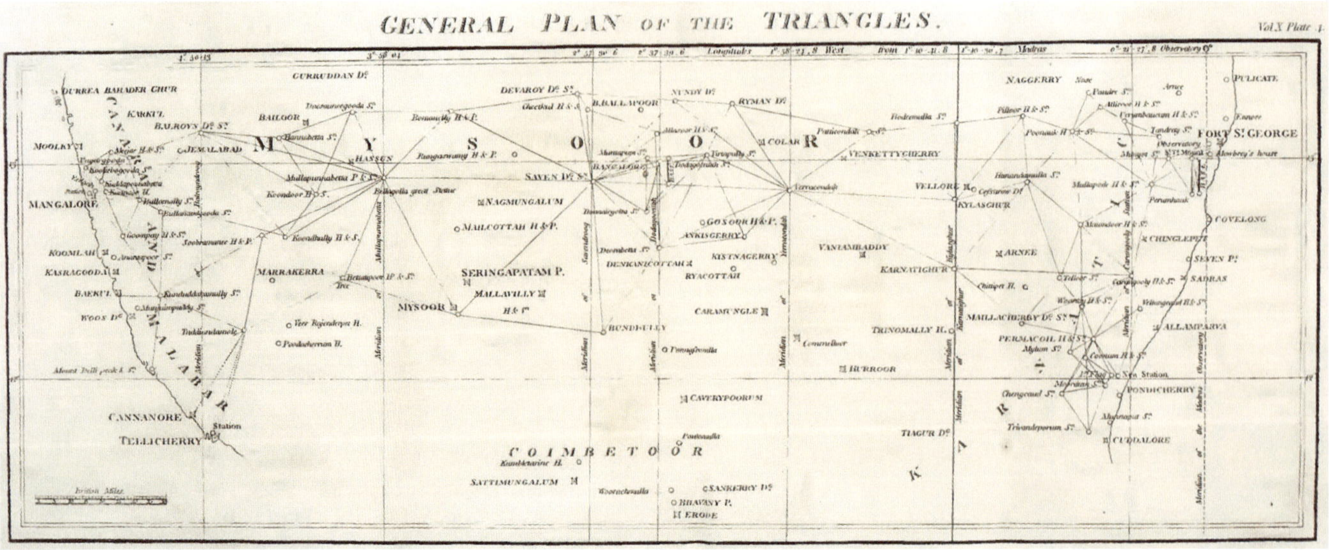

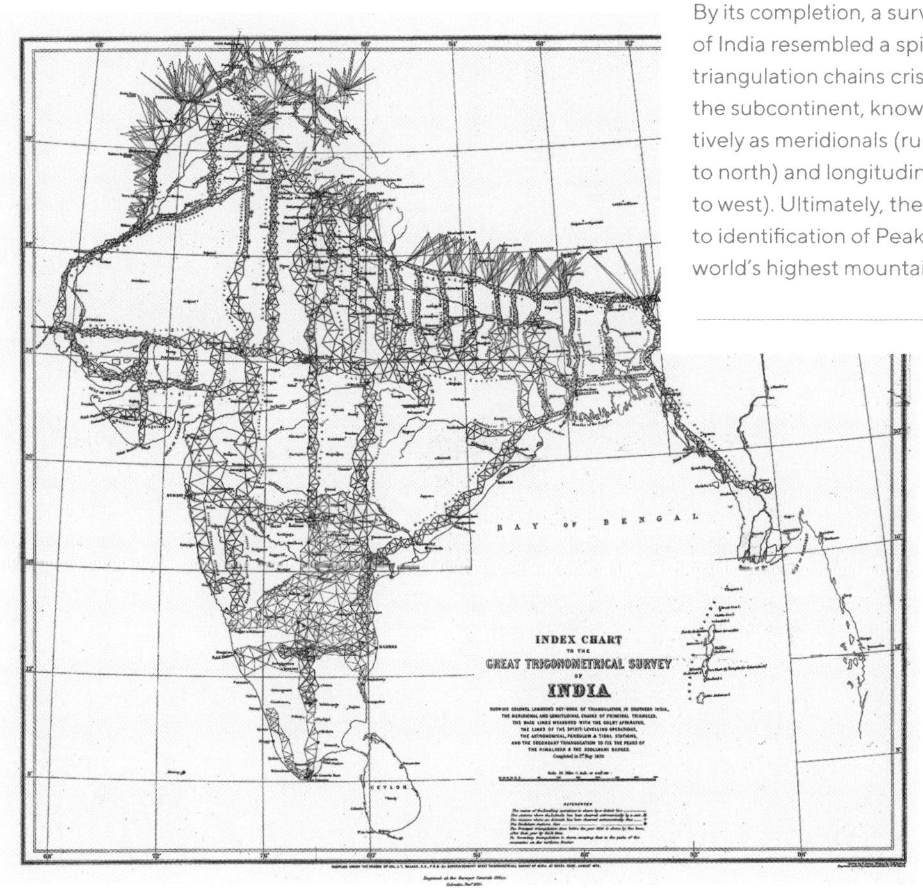

The Great Trigonometrical Survey of India was a mammoth undertaking. By its completion, a surveyor's map of India resembled a spider's web of triangulation chains criss-crossing the subcontinent, known respectively as meridionals (running south to north) and longitudinals (east to west). Ultimately, the GTS led to identification of Peak XV as the world's highest mountain.

Eventually another mountain top caught the eye. It was in 1849, after painstaking measurements from a number of different locations, that the so-called Peak xv was calculated at being over 29,000 feet (8,800 metres) and thus the highest peak in the Himalayas.[2] Despite George Everest's objections and in contradiction of the British preference for preserving local names, Peak xv was subsequently given a title in recognition of Waugh's predecessor as head of the Survey of India. Ever since, the mountain has been almost universally known as Mount Everest.[3]

Despite the excitement of establishing Mount Everest as the world's highest peak, its topography would remain veiled in mystery until the first reconnaissance mission to the mountain in 1921. The first map of Everest was largely the work of Canadian geographer and mountaineer Oliver Wheeler who was working for the Survey of India and was deployed on the expedition as surveyor.[4] To round the circle, Wheeler himself became Surveyor General of India in 1941.

Endnotes

1. Kangchenjunga has been measured at 8,586 metres, Nanda Devi at 7,816 metres, and Dhaulagiri at 8,167 metres.
2. According to the latest official measurements, it is 8,848.86 metres high
3. The Nepalese name for Mount Everest *Sagarmāthā* and the Tibetan *Chomolungma.*
4. See chapter 15.

Map showing part of Kishen Singh's legendary four-year mission (1879-1882) to explore Tibet and Mongolia. It proved a tale of endurance beyond compare. Working as a so-called 'pundit', Singh and his compatriots - working for the British - performed deeds of epic proportions which perhaps outmatched the achievements of more renowned adventurers of the day.

Detail, see full map on page 131.

The steppe-counters

Acting in the service of the British, native agents in disguise were sent on clandestine missions to map forbidden lands beyond the Himalayas

Whereas nowadays, with our mobile phones at our fingertips, step-counting and location-tracking are taken for granted, over 150 years ago far more rudimentary techniques based on the same principles were being employed on the Indian sub-continent to map unknown terrain across the mountains. Sent out on clandestine missions to cover vast tracts of hostile territory beyond the sphere of British influence, the surveyors came back – often after years of impoverishment and destitution – with data which was worth more than its weight in gold for the cartographers of the day.

By the time George Everest retired in 1843, the Great Trigonometrical Survey's task of measuring India from south to north along the Great Indian Arc of Meridian was largely complete. In the decades that followed, the Survey dedicated itself to covering the rest of British India with a mesh of secondary meridionals and longitudinals which criss-crossed the sub-continent (see chapter 11). As the 19th century progressed, gaining an improved knowledge of the northern extremities of British-controlled territories became the priority, primarily because lands beyond the Himalayas were yielding to the influence of Tsarist Russia. The belief was – real or imaginary – that the tsar harboured designs on invading India from the north.

Without ever coming to blows, a game of cat and mouse for geopolitical strategists was played out for the rest of the century by two superpowers vying for influence over neighbouring states. This, an early form of 'cold war', became known in English as the Great Game and in Russian, perhaps more evocatively, as the Tournament of Shadows. As part of its stratagem, it was vital for Britain to stake out the outer reaches of its empire so it could gain a more precise knowledge of what lay beyond.

Surveying the topography on the Himalayan fringes, however, was proving a bit of a headache to the British. Nepal, in the east, with the lion's share of the mountain giants, was a fiercely independent kingdom which pursued a strictly isolationist policy and sealed off its borders to Europeans. Kashmir, in the north-west, on the other hand, had been conquered by the Sikh Empire in 1819, with its Maharaja rulers strategically aligning themselves with the British. One benefit of this favoured status was the granting of permission to have its territories surveyed, the northernmost confines of which included the second highest mountain range in the Himalayas, the Karakoram.

Captain Thomas Montgomerie, a promising young Scottish surveyor working for the Great Trigonometric Survey (GTS), was appointed head of the Kashmir Survey in 1855. At the time, there were still lingering doubts as to whether Mount Everest - whose height had been fixed at 29,002 feet (8,840 metres) using trigonometric calculations from stations in India - was superseded in height by any of the Karakoram summits. Over the next ten years, under the guidance of Montgomerie, the GTS made significant advances in the mapping of the region. Its team of surveyors included William H. Johnson and Henry H. Godwin-Austen, who would both become accomplished mountaineers in their own right.

Johnson had a natural aptitude for working at high altitudes and was responsible for setting up triangulation stations at between 4,500 and 6,000 metres, and for more than sixty years, four of Johnson's peaks held the record for being the highest trig stations in the world.[1] Primary triangulation, which involved establishing hill stations at heights which had never previously been scaled, was complemented by plane-table surveying, which filled in the triangles with the intervening topography. Slowly but surely, the heights of the loftiest peaks in the Karakoram were logged. British surveyors started numbering these, from K1 to K32, their local names thus far not having been ascertained. K2, the highest of these was eventually recorded at 28,250 feet (8,611 metres), confirming it as the world's second highest mountain.[2]

Although the survey of Kashmir set a new benchmark in high-mountain mapping, with the threat from an expanding Russian empire on its northern frontiers, the British were desperate to know more about the terrain that lay beyond the Himalayas. To varying degrees of success, European travellers – sponsored or otherwise – had ventured north from India across the mountain passes in the preceding decades, some out of sheer curiosity, others to sound out the possibility of establishing trade relations with the khanates and sultanates of Central Asia. It was often a perilous undertaking: not only was the terrain extremely inhospitable, travellers going over the high mountain passes might also be preyed upon by brigands.

Even when these hurdles had been negotiated, the jurisdictions through which they passed were either highly suspicious of, or worse still, downright hostile to foreign intruders. To explore, alien travellers had to 'go local' and disguise

A group portrait of Great Trigonometrical Survey of India staff, together with a theodolite, in c. 1857.The four figures are (left to right): Thomas George Montgomerie, Sir Andrew Scott Waugh, James Thomas Walker and Sir Henry Edward Landor Thuillier. Montgomerie was head of the Kashmir survey, whilst Waugh, Walker and Thuiller all became Surveyor General of India at some point.

and conduct themselves as itinerant traders or holy men – not an easy task for Europeans and one that was full of risk. Early adventurers, if they ever returned, may have brought back extraordinary tales of exploits in exotic lands, but they had little to show in the way of accurate maps.

At most, knowledge of the mountainous topography was sketchy and incidental. By the middle of the 19th century, the need for Britain to extend its knowledge of the uncharted regions of Central Asia through more systematic exploration and surveying became more pressing. This need was underlined by intelligence reports of Tsarist Russia currying favour with a closed, inward-looking Tibet and expanding its geopolitical influence further south. It was a menace that Britain felt had to be thwarted, but Tibet, with its hermetically sealed borders, barred all access to foreigners. Without maps the country was a vast and empty canvas.

Montgomerie came up with a novel idea to address this problem. Although it was downright dangerous for Europeans to cross into territories which were ostensibly out-of-bounds, he had observed that locals could travel virtually unchecked into these neighbouring territories, often as traders in caravans to the far-off cities of Central Asia. What, he asked himself, if some of these natives could be trained as surveyors and sent across the frontier to explore on behalf of the GTS? In 1863, with the backing of his superiors at the GTS, his plan received the approval of the government, who saw this as a less hazardous undertaking than sending British officers into the void.

The plan was to recruit locals who would receive training in route-surveying and reconnaissance at the headquarters of the GTS in Dehra Dun, northern India. First and foremost, over a period of a year, they would be rigorously coached to walk at a standard pace over long distances and over different terrain. Before embarking

The 1864 map of the the Mustakh Range in the Karakoram. As part of his work on the survey Henry Godwin-Austen fixed the height and position of K2 (on the far right of the map) for the first time.

Sketch Map of the
USTAKH RANGE (TRANS-INDUS)
and
EY OF SKARDO &c.
by Captn. H.H. Godwin-Austen.
English Miles.
10
15
20
umps marked thus .... c.
76°
76° 30′
36°
35° 30′
YARKUND
YARKUND OR KARAKASH
KARAKORAM RANGE
SALTORO
Nobündi Sobündi
Drenmüng
Skeenmang
Tsokar
Chiring
Shing chukpi
Punmah Glacier
Chongulter
Punmah
Tsok
Korophon
Biaho Gl.
Biaho R.
Baltoro Gances
Monibrause
Tsokar
Shaktok
Spantok
MUSTAKH (pass)
about 18400 ft.
K.2. 28.265 ft.
K.3.
Güsherbrüm
K.3a
Masherbrum
25000 ft.
Masherbrum Gl.
Atoser Gl.
Nang Glacier
Hüshe
Zoah
21000 ft.
Hüshe R.
K.6. Large Glacier
Mingiloo
Thallé Valley
Chungokut Go.
Konkon
Monlong
Kerko
Kiris
Kunis
Bara
Tanskia
Kurphak
NUBRA R.
Kapaloo
Edwd. Weller.
by J. Murray, Albemarle Str. London, 1864.

on a mission into unknown territory, recruits would be tested along routes in India which had already been mapped as proof of their competence. But once they had embarked on a mission outside British-controlled territory north of the Himalayas, they would be on their own. To make them inconspicuous as they travelled, they would disguise themselves, for example, as Buddhist pilgrims who, in addition to their personal belongings, would carry concealed instrumentation. Their 'step count', for example, would be fastidiously kept using an adapted string of Tibetan prayer beads. Along the route they would note down any important topographic features and other information that would be useful to British intelligence. At the end of each day, these notes would be concealed in a prayer wheel which was designed for this purpose. In addition, they were instructed in the workings of a sextant and compass, learned how to identify stars and use these for locating positions. They were also issued with a boiling-point thermometer, which would help them determine altitude. As time went on, these instruments and the methods of concealing them became ever more sophisticated.

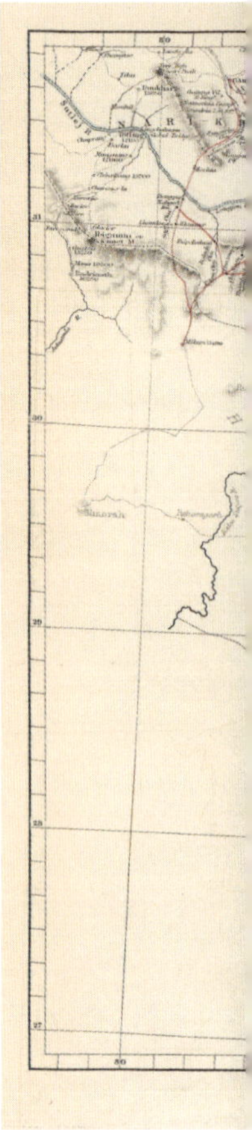

To all intents and purposes, these explorers were secret agents acting on behalf of a malevolent foreign power. If they were ever caught, they would have no redress to their employers, their deeds considered to be a misdemeanour punishable by no less than death. Once they had left British India, they were left to their own devices, relying solely on their intelligence and ingenuity for their survival.

Nain Singh and his cousin Mani Singh from the village of Milam in the Kumaon Himalayas in northern India were amongst the first of the so-called 'Pundits', the name given to local surveyors working on behalf of the GTS. The two cousins were chosen by Montgomerie to carry out the first exploration of Tibet and they had impressive credentials: Nain Singh spoke fluent Tibetan and was highly familiar with the customs of that land. Both he and Mani had earned their spurs on expeditions undertaken by the German Schlagintweit brothers[3] in India and Central Asia between 1855 and 1857. Once they had been selected, they were given training in route surveying at Dehra Dun.

In 1865 they were assigned the task of reaching Lhasa, the forbidden capital of a forbidden country. In their first efforts to cross the frontier through Nepal they were stopped and sent back by suspicious border guards. They retraced their steps back to the Nepalese capital Kathmandu, where the two cousins parted company. Mani returned to Dehra Dun whilst Nain Singh disguised himself as a Ladakhi merchant, complete with pigtail, and joined a trading caravan bound for Lake Manasarovar in Tibet, close to where the country's borders with Nepal and India meet.

On his way to Lhasa, on crossing the River Tsangpo, he was told the water flowed eastwards before turning sharply south into India, something that was news to Western geographers.[4] Striking north-east upstream along the Kyi Chu river, a tributary of the Tsangpo, he managed to reach Lhasa in January 1866. There, he rented a room, explored the city and took scientific observations. During his three-month sojourn, Nain Singh even managed an audience with the Dalai Lama. For extra funds, he spent time teaching, where his cover was almost blown after his

Nain Singh's expedition from Nepal to Lhasa (1865-1866).

disguise had been spotted by two Kashmiris. Eventually, he returned by way of western Tibet where he crossed the Himalaya back into India.

After a year and a half of topographical observations, Nain Singh had completed the first ever survey of Lhasa and the main trading route into Tibet from India. It was an extraordinarily successful mission. During his expedition, he established Lhasa's exact position by way of celestial observations and he used boiling water to estimate its altitude at 3,420 metres – 100 metres short of its actual elevation. He mapped the course of one section of the Tsangpo and brought with him climatic, topographical and political details which proved of immense value to the Survey. In no small part, it was due to the achievements of Nain Singh that more funding became available for future forays by pundits into Central Asia.

If Nain Singh was one of the trailblazing pundits, his younger cousin Kishen Singh – born in the same village – was surely the greatest. He was 16 years old when he started training at Dehra Dun in 1867. By the time he embarked on his most expansive voyage eleven years later, he had already covered over 3,000 kilometres on long-distance sorties into Tibet and neighbouring regions in the service of the Survey of India.[5] In 1878, Montgomerie's successor James Walker gave him instructions to travel by way of Lhasa across the vast expanses of Tibet into Mongolia. By their very nature, route surveys were unpredictable, but Kishen Singh's four-year mission was to prove a tale of endurance – and an ordeal beyond compare.

In April 1878, Kishen Singh set out from Darjeeling in north-eastern India on a trek into the unknown with his two companions-cum-servants, Chhumbel and Ganga Ram. In early September they arrived in Lhasa. Held up in the Tibetan capital for a whole year waiting for a departing caravan to join, the resourceful Singh occupied himself, amongst other things, by making a detailed survey of the city and learning Mongolian.

To reach the northernmost post of their journey at Dunhuang, 1,600 kilometres north of Lhasa, on the edge of the Gobi Desert, their caravan had to cross the immense Chang Tang plateau, on which the head-streams of several major Asian river systems, such as the Salween, Mekong and Yangtze rise. It was a merciless expanse: apart from a few nomads and marauding bandits, this vast area was uninhabited. Along the way, Chhumbel lost a toe to frostbite and their animals succumbed to cold and hunger. Beyond the Kunlun Shan, the party was then set upon by a group of over 200 brigands and in the engagement, Kishen Singh was robbed of all his possessions, save for his precious instruments. Having found work with various employers as they trekked slowly north, Kishen Singh also had to suffer the ignominy of being robbed and deserted by one of his own: Ganga Ram, his servant.

The inscription on the photograph reads: *Nain-Singh, chief Pundit trained by colonel Montgomerie, survey of India; Kishan Singh, asst. Pundit, the explorer of the Tengri-Nur Lake of Lhassa; an ordinary lama or Buddhist monk; a Tibetan dog of the small terrier kind.*
Date: 1850-1869

It was not until January 1881, now almost three years into their expedition, that they reached Dunhuang, an oasis settlement on the ancient Silk Road famous for its sacred Buddhist caves. No sooner had they arrived, than they were detained as suspected spies by the city's governor, their greatest luck perhaps being the failure on the authority's part to discover their surveying equipment. After another seven months, their confinement only ended when they were released on the request of a Tibetan lama, who promised to take on Kishen Singh and Chhumbel as servants on a journey to his monastery home 960 kilometres to the south-east in Tachienlu.[6] It was there, to their good fortune, that – via Jesuit priests who had a mission there – they were able to send a message back to Dehra Dun, telling General Walker of their survival.

Two months later they were released and then set about making their way back to India from this, the easternmost point of their marathon survey. First they had

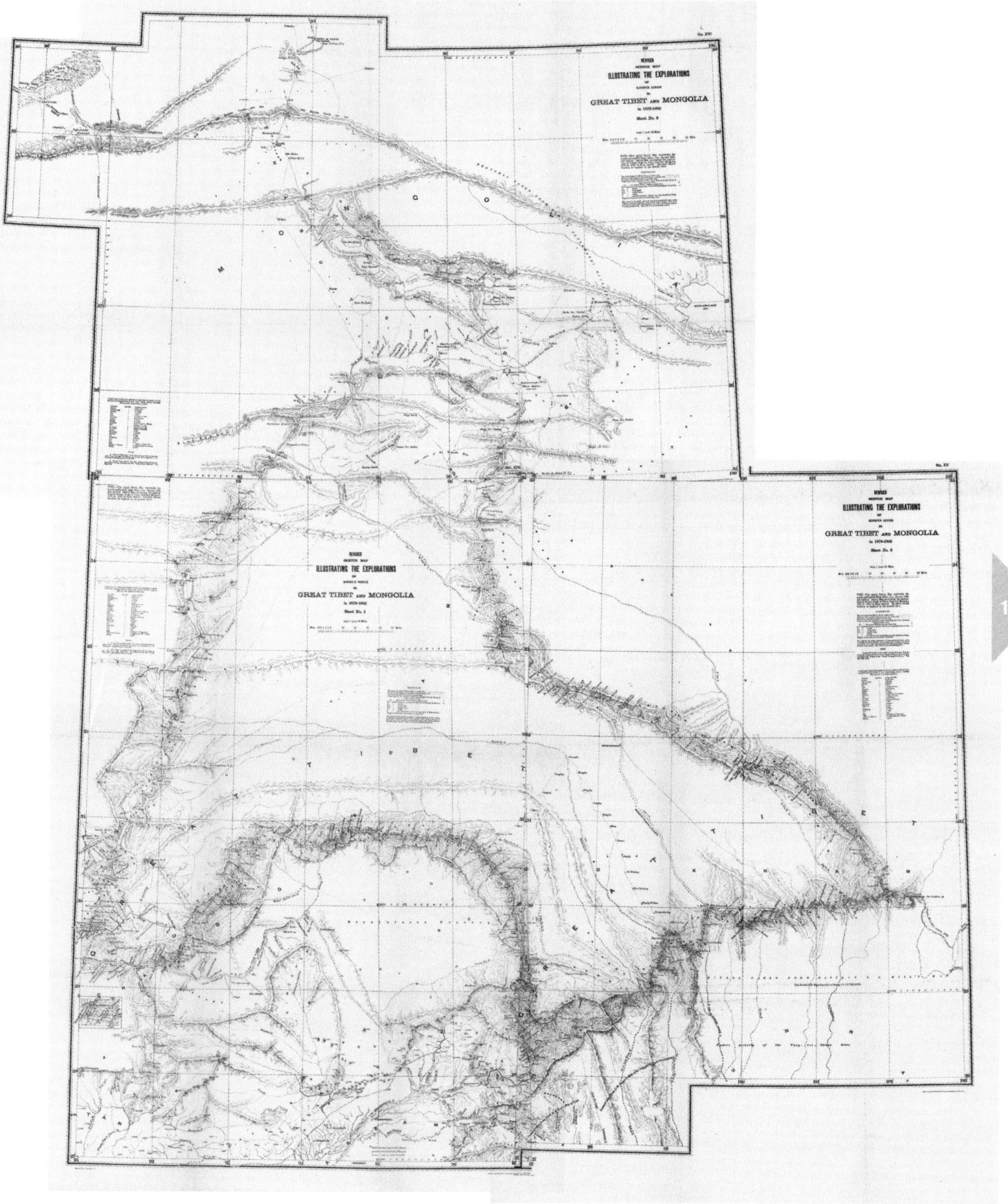

Kishen Singh returned from his mammoth journey through Tibet and Mongolia with all his survey equipment intact and his survey notes complete. In all, he and his companion Chhumbel clocked up a staggering 4,500 kilometres, or five million paces.

to navigate their way through perilous 'Deep Gorge Country' where three of the great rivers of Asia – the Yangtze, Mekong and Salween – run parallel to each other, north to south, through steep ravines, bordered by 6,000-metre-high mountains. After successfully traversing the demanding terrain, they ended up in a local outbreak of smallpox and were held up by twenty days of quarantine. Once released from this confinement, they found themselves just 50 kilometres from Assam in British India. Though lopping off a considerable amount of mileage, this short-cut would however mean passing through lands populated by the Mishmi people, who had had a treacherous reputation: in the recent past they had murdered French missionaries after promising safe passage through their country. Fearing for their lives, they opted for the safer but longer route westwards through Tibet and thence to northern India. The upside for the Survey was that they would cover yet another 1,000 kilometres of unmapped territory.

It was now May 1882. They were almost back on familiar ground and close to home, yet were subject to further misadventure: shortly after passing a Tibetan penal colony they were seized on suspicion of having escaped, not least due to their shoddy exteriors, and were only released on the chance appearance of an acquaintance who vouched for their innocence. It was not until July that they were allowed to resume their homeward journey, the mountain passes not being clear of snow as a result of an unusually late spring. While they waited for the passes to clear, Kishen Singh went from house to house in the neighbourhood reciting Tibetan sacred books in an effort to replenish their much depleted funds.

Finally, on 12 November 1882, four and a half years after setting out, they arrived back in Darjeeling. According to General Walker, *"they arrived in a condition bordering on destitution, their funds exhausted, their clothes in rags, and their bodies emaciated with the hardships and deprivations they had undergone."* Despite everything that had transpired on their mammoth journey, Kishen Singh returned with all his survey equipment intact and his survey notes complete. In all, he and his companion Chhumbel had clocked up a staggering 4,500 kilometres, or five million paces, mapping huge, unknown tracts of Central Asia. Singh had taken observations for latitude at twenty-two points and calculated elevations at seventy places, using the boiling-point thermometer.

According to the official report, Kishen Singh's figures produced results that were as creditable as those of contemporary explorers who were otherwise not subjected to the trials and tribulations of our heroic pundit.

The two cousins were awarded medals by the Royal Geographic Society for their services and given pensions by the Survey of India sufficient to live out a comfortable life in their homeland. From a modern-day perspective, they would surely be more prominently acclaimed. Thankfully, contemporary narratives refer to Nain Singh and Kishen Singh as explorers, a more logical description than pundit alone. Not only were they learned persons in this, the Hindi sense of the word, they performed deeds of epic proportions which are in every way comparable to the more renowned European adventurers of the day.

The 1906 map of Tibet and the Surrounding Regions by The Royal Geographical Society.

Endnotes

1 It should be remembered that these triangulation stations were significantly higher than those in the Alps during the Golden Age of mountaineering.

2 For a period of time, K2 was rechristened Mount Godwin-Austen, in honour of the explorer who had surveyed the immediate environs of the mountain. The name never stuck even though Godwin-Austen might be said to have a much stronger claim to having this peak named after him than George Everest, whose name was given to the world's highest peak - despite Everest's own protestations to this effect.

3 The Schlagintweit Brothers (Adolph, Hermann and Robert) were German explorers mostly remembered for their travels and research in India and the Himalayas between 1854 and 1858. In August 1857, Adolph, travelling alone in Turkestan, was suspected of being a Chinese spy and without benefit of a trial was beheaded on the orders of the Emir of Kashgar.

4 Geographers would discover that the Tsangpo, the longest river in Tibet, breaks through the Himalayas in a series of great gorges to form the Brahmaputra and then empties via the Ganges Delta into the Bay of Bengal.

5 Under Montgomerie's successor James Walker, the Great Trigonometrical Survey was amalgamated into the Survey of India.

6 In Chinese, Kangding

Alpine clubs and mapping

Until the middle of the 19th century, mapping was a domain that was almost exclusively in the hands of the military. It was around this time however, that tourism and, more specifically, a growing popularity for mountaineering, generated a demand for a more detailed knowledge of the high mountains. Practitioners of the sport began to organise themselves into mountaineering clubs, the first of which was the Alpine Club, founded in London in 1857. This was followed by a spate of other clubs being formed between 1862 and 1874, namely the *Österreichischer Alpenverein* in Austria, the *Schweizer Alpen-Club/ Club Alpin Suisse* in *Switzerland,* the *Club Alpino Italiano* in Italy, the *Deutscher Alpenverein* in Germany and the *Club Alpin Français* in France.

With the exception of the Alpine Club, mapping became one of the priorities of these associations, the reason being that official maps at the time failed to meet the needs of mountaineers. The Swiss Alpine Club[1] and the *Club Alpin Français* only published maps for a limited period before nationally recognised mapping organisations satisfied the requirements.

This was different in Germany and Austria, where the Alpine clubs took up mountain cartography with a vigour, especially when the two clubs were amalgamated as the *Deutscher und* Österreichischer *Alpenverein* (DuÖAV)in 1874[2]. At first, maps were published with different scales and designs, but at the turn of the century, the club started producing maps of the main mountain regions in the Eastern Alps at a scale of 1:25,000. In fact, almost all sheets that were produced during this period can still be obtained in their modern-day versions today, having been subject to regular updates over the last century. As well as conventional methods, in 1922 the DuÖAV introduced the pioneering technique of photogrammetry in its standard cartographic toolbox. This is a surveying method which compares information from a series of photographs – terrestrial or aerial – and uses stereoscopic plotting instruments to produce accurate mapping. One of its benefits was that it enabled cartographers to plot contours more precisely. Since 1935, the club has also been producing a number of maps of mountain regions outside Europe, most notably in the Andes and the Himalayas.[3]

Other Alpine clubs may no longer be producing maps of their own, but they still provide a rich source of historical material. For example, some years ago, the library of the *Club Alpino Italiano*, based in Turin, undertook a project to digitise the historical maps it holds in its archives. So far over 1,000 maps have been digitised and made available online for geographical research, to study the evolution of mountain topography, or simply to appreciate the sheer beauty of mountain maps. A number of historical maps from their collection have been used for this publication.

The 1:66,000 map of the Glockner group (in present-day Austria) by Peter Wiedenmann, from the 1871 journal of the German Alpine Club.

Terrainstich v. H. Rohn

Herausgegeben v. Deutschen u. Oesterreichischen Alpen Verein
1903
Kartogr. Anst. G. Freytag & Berndt Wien

The 1:25,000 map of the Brenta Dolomites in Trentino published by the German and Austrian Alpine Club (DuÖAV) in 1908. Up until 1919, Trentino was part of the Austro-Hungarian Empire, but it was ceded to Italy after the First World War. Cima Brenta, the highest peak in the range (3,152 m) is also marked at the very top of the map as Kaiser Franz-Joseph Spitze. The German and Alpine Alpine Clubs continue to map the Dolomites (now wholly within Italy) to the present day.

Endnotes

1 See separate sections on Swiss cartography: Swiss style: *snapshots from a nation's cartographic past*

2 After the Second World War, the Austrian and German Alpine Clubs once again became separate entities – the ÖAV and DAV, but have continued to collaborate in the field of mapping.

3 Richard Finsterwalder was a leading geodesist and cartographer who worked extensively for the German Alpine Club and was responsible for producing the first detailed maps of the Fedchenko Glacier for the 1928 Russo-German Alai-Pamir Expedition (see chapter 16) and Nanga Parbat for the 1935 German Nanga Parbat expedition.

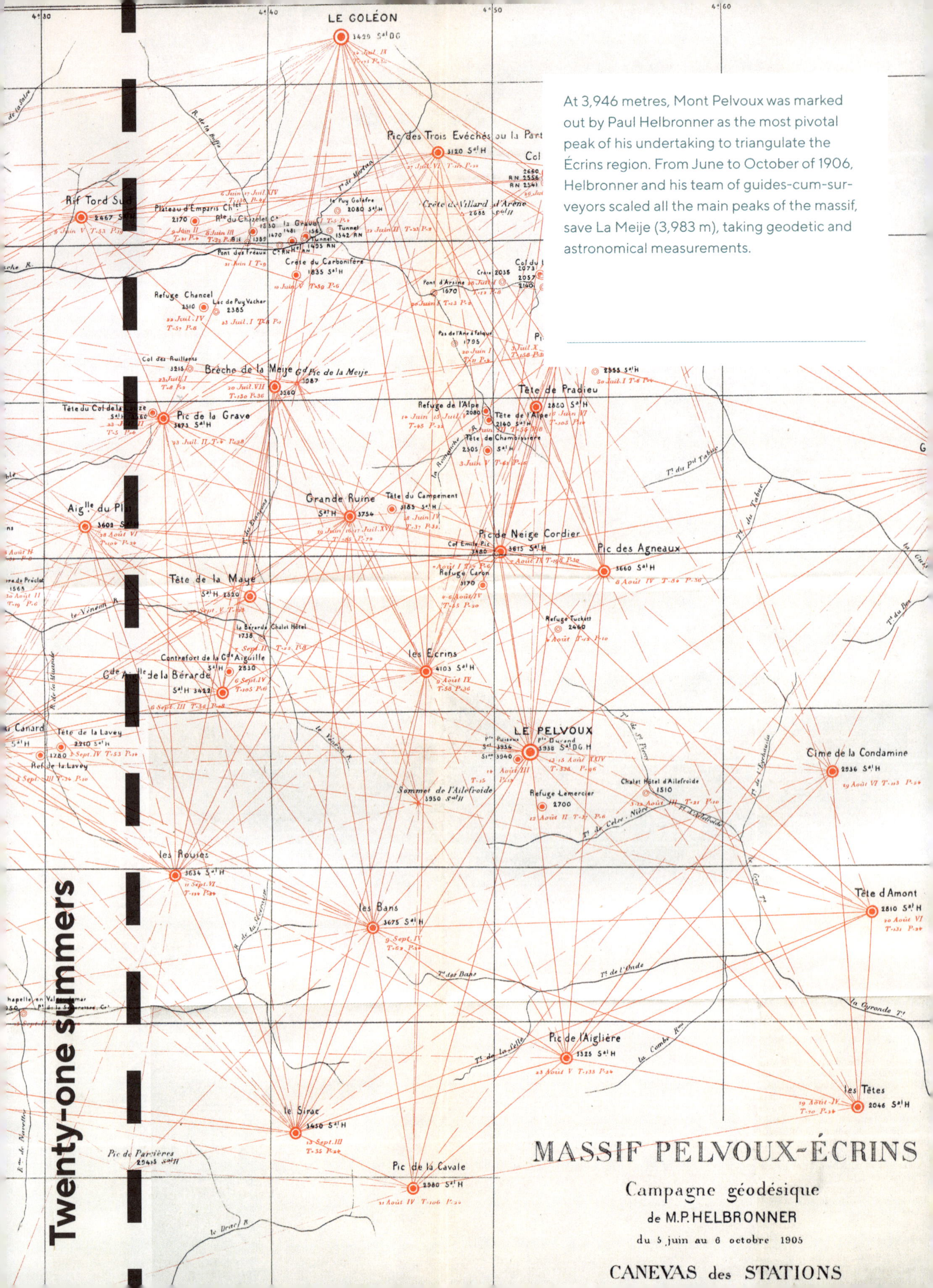

At 3,946 metres, Mont Pelvoux was marked out by Paul Helbronner as the most pivotal peak of his undertaking to triangulate the Écrins region. From June to October of 1906, Helbronner and his team of guides-cum-surveyors scaled all the main peaks of the massif, save La Meije (3,983 m), taking geodetic and astronomical measurements.

Twenty-one summers

After a summer spent on Mont Blanc in his youth, Paul Helbronner spent his adult life in pursuit of a dream to map the French Alps

Stepping out of the futuristic Sky Monte Bianco cable car at 3,462 metres, few amongst the multitudes of day-trippers ascending from Courmayeur today will have any notion of how Pointe Helbronner – on which the cable-car station is perched - came by its name. The first clue is in the uninterrupted 360-degree view from the circular terrace from where, the website proclaims, *"you can almost reach out and touch the mountains all around"*. The sweeping views across the high mountain terrain afforded by this isolated peak provided the perfect vantage point for a young surveyor exploring Mont Blanc.

Little did the youthful Paul Helbronner know when he started exploring the Mont Blanc massif that this once isolated rock pinnacle would one day bear his name and ironically see tens of thousands of tourists on its summit every year. Yet his upbringing did not suggest his destiny was to spend much of his adult life obsessively surveying the French Alps. Helbronner was born into a wealthy Parisian family in 1871, which moved within the higher echelons of French Society. He came from impressive stock: grandfather Helbronner, originally from Bavaria, managed a design house in London which supplied soft furnishings for chic clients, including the Princess of Wales. Paul's mother was an artist and his father a barrister, his best man at their wedding was Jules Grévy, a future president of France. Following his father's untimely death at the age of 37 when Paul was just nine years old, the most important influence on his formative years became his uncle, Auguste Michel-Lévy, a geologist who had helped compile the geological map of France in 1876. With this kind of pedigree it was no wonder that he became a polymath at an early age, excelling at a variety of academic subjects, including history and geography, and sports, in particular horse-riding and fencing. As a teenager he also took an avid interest in photography. It was not until the summer of 1883 however, that he had his first taste of the mountains when he went on a two-month stay in the Central Massif with his uncle.

In 1889, Helbronner was suffering from growing pains which led to migraines and his Uncle Auguste decided to take him to Chamonix to recuperate. They lodged at the Hotel Couttet, a favourite haunt for mountaineers, founded in 1862 for the climbing community.[1] From here, Paul explored the mountains and learned the rudiments of climbing and was instantly hooked. Two years later, on the back of his training as an alpinist, he made his first ascent of Mont Blanc. On each occasion he headed into the mountains, his photographic equipment went with him and his experiences were methodically penned in a journal he kept.

Photograph of Paul Helbronner at Pointe Durand, one of the two summits of Mont Pelvoux in 1921.

Given the circles within which he moved, it is probably no coincidence that during a stay in Chamonix in 1892 he was introduced to Joseph Vallot, a geodesist and mountaineer. At the time, Vallot was on a scientific mission on Mont Blanc, working from an observatory he had constructed close to the summit, at 4,362 metres. Part of the scientific work involved producing a detailed map of the massif at a scale of 1:20,000, triangulation for which was being carried out by his cousin, Henri Vallot.[2] The encounter with the Vallots that summer inspired Helbronner to enrol at the École Polytechnique in Paris, a leading academy sponsored by the armed forces, which allowed him to pursue studies in mathematics, astronomy and geodesy. Helbronner was invited to join the Vallots on the mountain the next summer[3], where he learned to hone his newly acquired skills as a surveyor and cultivate an aptitude for photography, most notably by taking panoramic views from prominent tops. Ultimately, it was these two talents that would prove to be his legacy.

Over the years, his relationship with the Vallots intensified and they would also regularly meet during the winter months, between their busy social engagements in Paris. Helbronner would use the 'close season' to develop the negatives he had brought back with him from the previous summer. His mother also encouraged him to refine his artistic flair by taking classes with the famous military artist, Édouard Detaille.

Aged 26, Paul Helbronner married Héléne Fould, heiress to the fortune of the Fould-Dupont steel manufacturing company in Nancy. In the years that followed, the distractions of helping run the business and the obligations of a family man with three young children may have kept Helbronner from the call of the mountains. His years away from the hills spurred him on to work on a series of vividly painted illustrations based on his mountain panoramas, each of which were elaborately labelled with the names of peaks and other topographic features.

Nevertheless, it was the seemingly inexhaustible financial resources he had acquired through marriage which would ultimately provide the launchpad for fulfilling his dreams and the next and longest chapter of his life.

The most defining moment of Helbronner's life came in 1902 when a proposal was put to the *Club Alpin Français* (CAF) to carry out a geodetic survey of the French Alps with the aim of creating a large-scale 1:40,000 map. The idea was to replicate the success of the Siegfried map of Switzerland[4] and was born out of a need for a more detailed representation of mountain topography, which at that point the national mapping agency, under the *Dépôt de la Guerre* with its military rationale, was unable to offer. Henri Vallot and Paul Helbronner took it upon themselves to set up a foundation (*Caisse d'action du Club en montagne*) to fund these efforts and the CAF voted unanimously in favour of their proposal in February 1903.

Helbronner would be in charge of its execution. A *plan de campagne* was worked out which would run from north to south. In the space of five months, Helbronner put together a formidable collection of equipment for taking geodetic and astronomical measurements from designated locations for the territory they planned to cover in the summer of 1903. In addition to the conventional use of the theodolite, modern techniques of photogrammetry and stereoscopy would be applied. Designs were drawn up for the signal stations that were to function as triangulation points, large enough to be visible from afar. It's clear from his copiously filled notebooks that Helbronner was meticulous in his planning. To take advantage of the best conditions, readings from each station would have to take place before ten o'clock in the morning and the higher the peak, the earlier the start. Because the time necessary for completing observations would be between five and seven hours, nights out on the mountains would not be unusual, especially if it was a part of the primary triangulation.

Work started in earnest on 18 July that year in the Puy Gris massif, rising to 2,908 metres in the northern part of the Chaîne de Belledonne near Grenoble. Helbronner's notes state that his work was plagued by bad weather, but in 22 days, operating from eleven primary or secondary stations, he was able to fix the location of around 85 trigonometric points in the mountain range.

The 1903 season was a mere foretaste of things to come. The survey report for that year already extended to two volumes, both seven centimetres thick. In correspondence between Helbronner and his mentor Henri Vallot, the former foresaw the task ahead as truly daunting, but Vallot had only praise for his efforts and offered every encouragement.

The account of the 1905 campaign in the Massif des Écrins is one of the most evocative. Not only is this massif considerably more extensive than that of Mont Blanc, it is the highest range outside of the Pennine Alps, culminating in Barre des Écrins at an altitude of 4,102 metres. At least ten of Helbronner's primary stations in Les Écrins were above 3,500 metres, so not only were additional manpower (in the form of guides) and equipment (including high-elevation tents) required, mountaineering prowess was an absolute prerequisite.

In the 1906 journal of the CAF, *La Montagne*, Helbronner provides a blow-by-blow account of this campaign in a report entitled 'Quatre moins de Triangulations dans le massif Pelvoux-Ecrins'. From June to October of that year, Helbronner and his team of guides-cum-surveyors scaled all the main peaks of the massif, save La Meije (3,983 m).[5]

At 3,946 metres, Mont Pelvoux was marked out by Helbronner as the most pivotal peak during the whole of the Écrins undertaking. On 13 August he set up camp on Pointe Durand, the slightly lower of the two summit prominences (Pointe Puiseux being fourteen metres higher), chosen because its rocky heights lent themselves more easily to the installation of his equipment. Perhaps too, it was in homage to Captain Durand, the surveyor who made the first ascent of the mountain in 1828 (see chapter 9). The fact that two of his heroes, Edward Whymper and Armand Durand both had strong associations with the mountain probably compounded his own sense of destiny. After two nights on the mountain, he woke up bleary-eyed in his tent on the morning of 15 August to be met by a vision of his fellow surveyor:

Suddenly, a man opens the two flaps of canvas forming the door of my abode ... I recognise him straightaway: Captain Durand, a fellow alumnus of the **École Polytechnique** *in 1808. ... We are neighbours, since he lives 8 metres below me... Immediately, our conversation turns to what fascinates us and [outside] he points out the stations which he had either just finished or still intended to do in the summer of 1830: the Pic de Bure, the Taillefer, the Aiguille du Goléon, the Thabor, the Rocher Blanc des Sept Laux, the Grand Veymont , the Perron des Encombres, the Pic du Frêne, the Mourrefret ... What astonishing courage [Durand had] to venture onto this peak [Mont Pelvoux] without even knowing if it is accessible!*

On 1 September that same summer, Helbronner made the ascent of Roche de la Muzelle (3,465 m), accompanied by the 73-year-old Pierre Gaspard, one of the most experienced guides in the district, perhaps most famous for the first ascent of La Meije. Indicative of the arduous task that his team faced in surveying the high mountains, this epic excursion into the wild surpassed even the exacting standards that Helbronner had set for himself. The party, which included 'Père' Gaspard, one of his sons and three other guides, had set off at 10.30 pm on the previous evening under the light of a lantern. After reaching the Lanchâtra glacier at 6.00 am the next morning, they had to negotiate an uncharted near-vertical wall to reach the main ridge along which the summit could be attained. Having to overcome a number of crags and clefts, Helbronner likened the traverse of this summit ridge to an exercise in gymnastics. Twelve and a half hours after departure, the party eventually arrived at the summit, where they spent another four hours on geodesic and photographic work. After an even more circuitous and hair-raising descent from the summit, 23 hours and 50 minutes later they eventually arrived back at their base.

Over the course of the next two decades, Helbronner continued his geodetic survey as far as the Côte d'Azur. However, not content to measure continental France alone, in 1925 Helbronner cast his sights south across the blue waters of

the Mediterranean. His bold plan was to establish a meridian arc of the island of Corsica (consisting of a straight line running between two fixed points on the same longitude) and join this (by way of triangulation) to the meridian arc of the Alps, which Helbronner had already measured. It would be no easy task: at its nearest point, Cap Corse, the island is 170 kilometres distant from the French mainland near Nice. Visibility over this distance depends on exceptional atmospheric conditions, usually at dawn or dusk and only during brief periods of the year.

Helbronner was, if nothing else, absolutely thorough in his preparations, eager not to make the same mistakes as those made during previous attempts to triangulate over such great distances.[6] From four stations close to the coast between Toulon and Nice and from stations on three of Corsica's highest peaks, he maximised the number of angles to twenty to increase the reliability of his readings. In Corsica, he created mule tracks to access the signals he had fixed on masonry pillars and built stone shelters to accommodate the survey team on prolonged summit stays. The makeshift signals he was to install and power had the same luminosity of a lighthouse – between 4 and 10 million candela, no less – so that these could be identified at night with a pinpoint accuracy from the Côte d'Azur. To help achieve his aims, he was granted generous support in terms of equipment and personnel by the Ministry of War and the Navy, on condition that no additional expenditure would be charged to the state. The rest Helbronner paid for from his own seemingly limitless reserves.

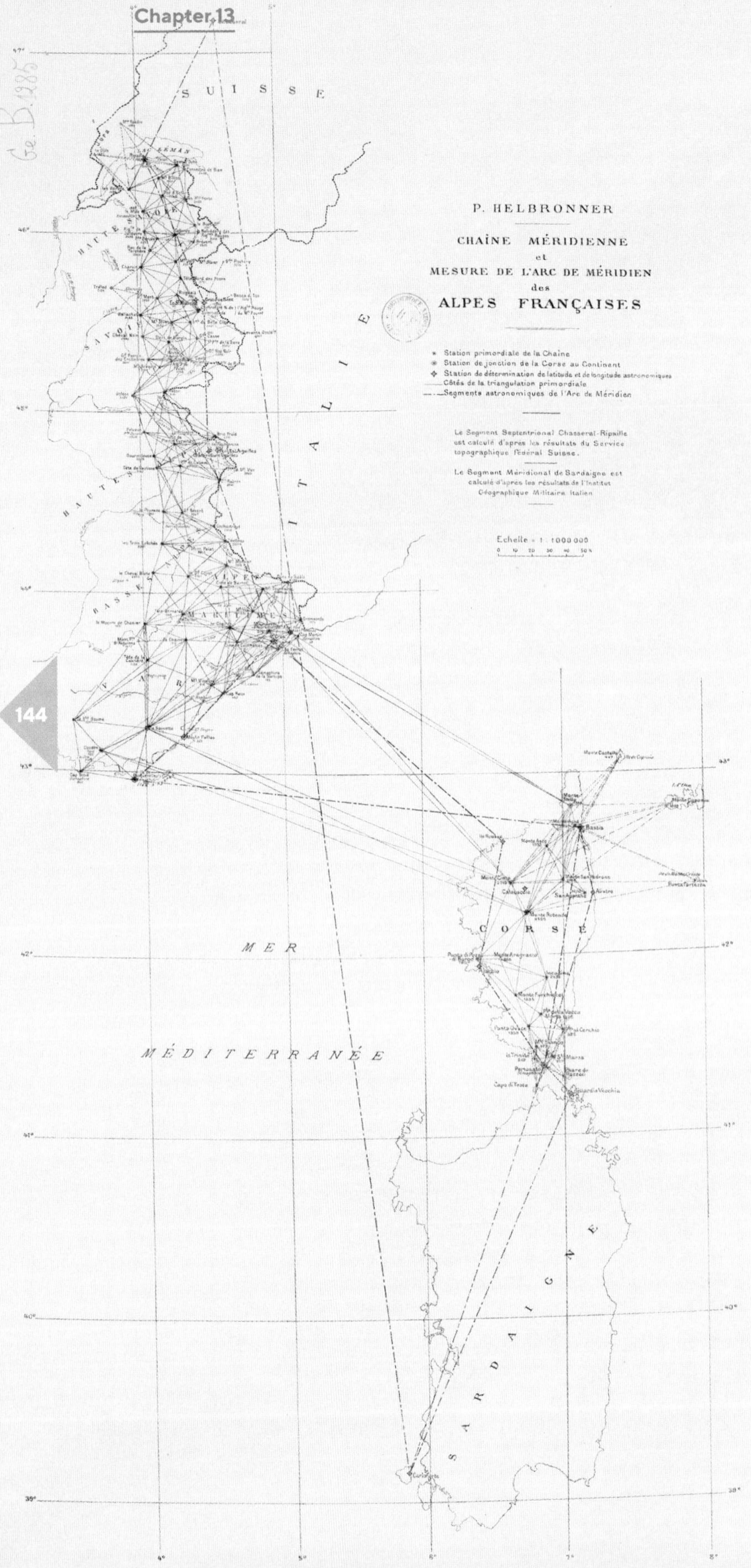

In 1925, Helbronner joined a meridian arc of the island of Corsica with the meridian arc of the Alps, which he had already measured. It was no easy task: at its nearest point, Cap Corse, the island is 170 kilometres distant from the French mainland near Nice.

During a four-month period from July until early November 1925, multiple readings were taken from either side of the Mediterranean divide, often involving several week-long sojourns in the summit shelters. On one particularly clear night, from the summit of Monte Cinto (2,706 metres), the highest peak on Corsica[7], it was possible to take uninterrupted measurements of the signal on Monte Stello on Cap Corse (70 km distant) and the mainland signals on Mont Agel (198 km), Mont Chauve (206 km), La Sauvette (236 km) and Mont Coudon (256 km). One reading taken from the Corsican summit of Monte Rontondo to Mont Coudon measured a staggering 271 kilometres, establishing a new record. In so doing, Helbronner accomplished his task of joining both meridians, from Geneva in the north to Corsica in the south.

Between 1903 and 1928, Helbronner spent 21 successive summers in the mountains, each season on average four months long. In 1910, having completed the arc of meridian from Lake Geneva to the Côte d'Azur, he published the first in his ten-volume series of works entitled *Description géométrique détaillée des Alpes françaises*, which presented his detailed geodesic measurements and computations (the last of these volumes was published posthumously in 1939). It was a truly astonishing achievement. He had calculated over 8,500 geodesic points covering 18,500 square kilometres, using triangulations from 1,818 stations, 151 of which were over 3,000 metres in altitude.

There are those who question Helbronner's legacy. By the time he had completed his work, his methods had largely been overtaken by advances in aviation and aerial photogrammetry, thus rendering his work extraneous, but this can hardly be the fault of a lack of foresight. He was first and foremost a geodesist, not a pioneering theoretician, but held a belief that his thousands of pages of measurements and geometric calculations were adding to a growing corpus of geodetic knowledge. At the same time he was desperate to receive acknowledgement from the wider scientific community for his achievements, often deflecting the efforts of others to his own advantage. Some accuse him of appropriating the ideas (and fortunes) of others for his own self-glorification: nowadays we might call this getting the lucky breaks in life. It has been suggested he was dictatorial in his approach: of course, he was a man who was single-minded, who went about his work with a missionary zeal and clearly not someone who was going to be easily blown off course.

Whether his work, from a scientific point of view, was ultimately in vain – after all, he left us no map or abiding theory, just a mass of calculations – is perhaps for the historians to decide. Notwithstanding, he left us over 15,500 photographic images, some of which he used as a source for his exquisite watercolour panoramas. And not insignificantly, in the copious notebooks, diaries and other writings, he left us extraordinary tales of adventure and endurance in the mountains. And there is an apposite symmetry to the fact that the only two locations on the IGN's current 1:25,000 topographic map of France which bear his name are at the two extremities of his meridian, namely Pointe Helbronner in the Mont Blanc massif and Abri Helbronner, the stone shelter he built on Monte Rontondo in Corsica.

Above: Helbronner left over 15,500 photographic images, some of which he used as a source for his exquisite watercolour panoramas such as these.
Below: Mont Pelvoux, p. 143: Mont Blanc

Endnotes

1 Edward Whymper was a frequent visitor to Hotel Couttet and died there in 1911. We know from Helbronner's journals from these years that he was an avid reader of Whymper's adventures in the Alps, so it was a stroke of luck that he should chance upon the famous English mountaineer in the garden of Hotel Couttet in 1893.

2 Helbronner established a life-long friendship with Henri Vallot.

3 It was probably during this period that the prominent pinnacle near the Col de Géant adopted the name Pointe Helbronner.

4 The Siegfried map of Switzerland represented an update of the Dufour 1:100,000 topographic map of Switzerland, the first official series of maps which encompassed the whole of the country. Named after Hermann Siegfried (1819-1879), the Chief of Staff of the Swiss Topographic Bureau, the maps were produced at a scale of 1:25,000 / 1:50,000 as the Topographic Atlas of Switzerland. Unlike with alpine clubs in other countries, the Swiss national mapping agency was able to fulfil the needs of its mountain-climbing members with these maps.

5 The main summit of La Meije, Grand Pic, was the last major peak in the Alps to be summitted in 1877.

6 Previous attempts to join Corsica geodetically with the mainland had been less successful. At the end of the 18th century, Jean-Joseph Tranchot (who later mapped the Rhineland - see chapter 10 - connected Corsica to Italy by triangulation, the precision of which was mediocre. In 1827, during operations for the *Carte d'État Major*, Capitaine Durand, having set up stations between Marseilles and Nice on the mainland carried out measurements connecting stations on two Corsican peaks over 200 kilometres distant, which were likewise too imprecise to be of any use. Other unsuccessful attempts were made by French and Italian surveys in the late 19th century.

7 According to the Wikipedia site on Monte Cinto, "Its location gives it a theoretical panorama of mountains on mainland Europe stretching from near Marseille to Rome. The most distant mountain theoretically visible is Monte Rosa in Italy, just west of north, approximately 405 km away." See also: https://www.viewfinderpanoramas.org/

P. Vittorio Emanuele
Campo V 4475
4901
M. Speke
Emanuele
4848 P. Johnston
4193
C. Stuhlmann
G. Johnston
4067 e
G. Alessandra
5125
P. Margherita
5105
G. Margherita
P. Alessandra
4805
Pianoro Stanley
M. Stanley
3933 d
Lago Bujuku
3748 g
3918
P. Moebius
P. Elena
4995
Campo IV
4515
Ghiacciaio Elena
4980
P. Savoia
Ghiacciaio Savoia
C. Scott Eliott
4347
Rª Grauer
4515
4654
4659
4347 l
n 4269
G. Semper
G. Baker
G. Moore
4349
Campo
4829
P. Semper
Campo III
4219
4174 a
P. Edoardo
4873
4118
4032
Campo
M. Baker
B
4101 m
3930 b
4071 c
Ghiacciaio Edoardo
4322
Campo VIII
4390 p
Campo II
4045
4326
C. Freshfield
A
P. Stairs
4590
G. Stairs
P. Sella
4659
G. Sella
G. Thomson
P. Weismann
4663
M. Luigi di Savoia

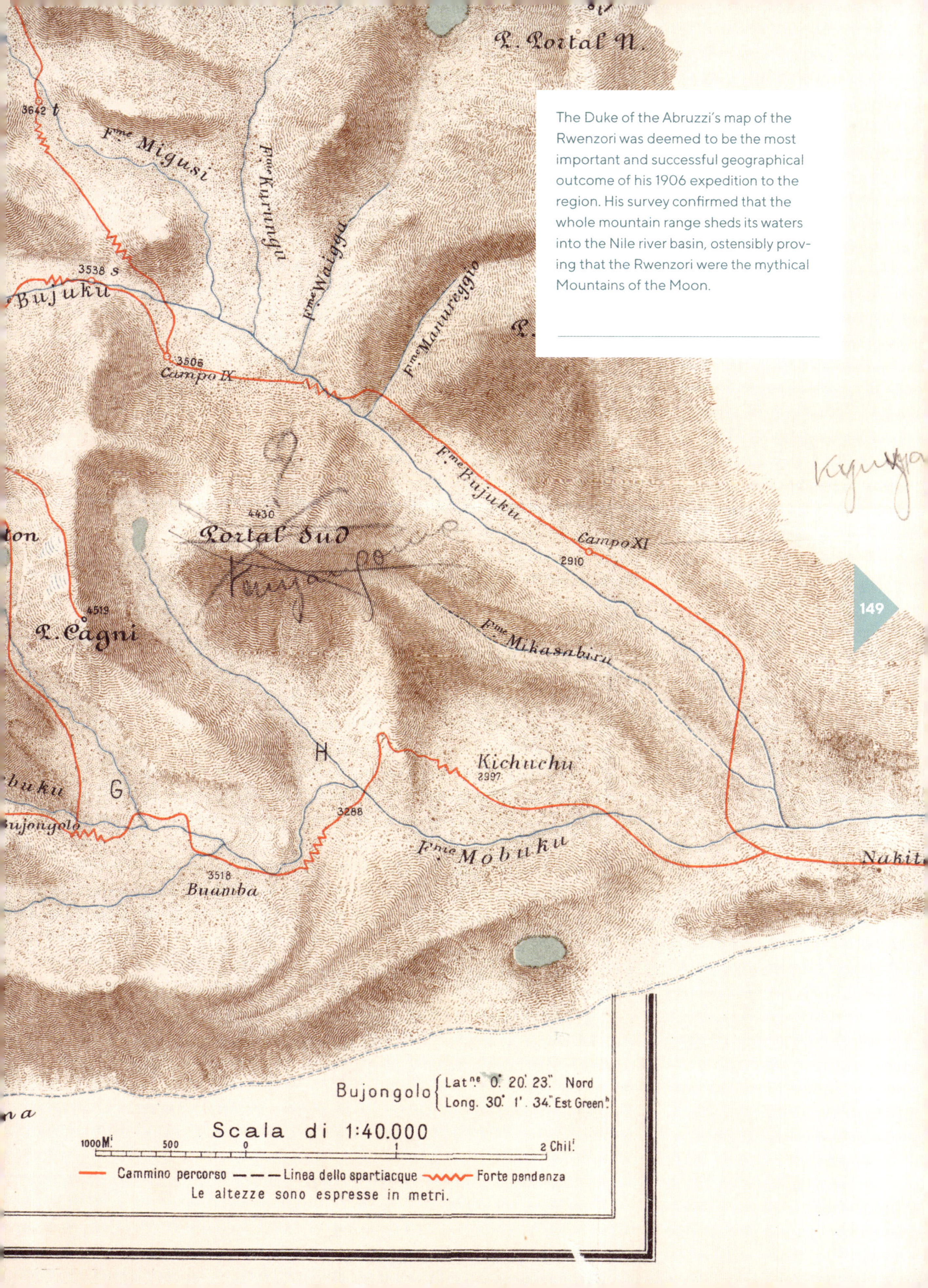

The Duke of the Abruzzi's map of the Rwenzori was deemed to be the most important and successful geographical outcome of his 1906 expedition to the region. His survey confirmed that the whole mountain range sheds its waters into the Nile river basin, ostensibly proving that the Rwenzori were the mythical Mountains of the Moon.

Chapter 14

To the Moon and back

A successful Italian expedition to a mysterious mountain range in the heart of Africa finally uncovered its secretive topography and equatorial snows

When Vittorio Sella returned to Italy in 1906 from an expedition to a barely known, snow-clad mountain range on the African equator, his dazzling photographs enraptured the general public back home in Europe. Doused in a thick blanket of mist for most of the year, the ice-topped mountain peaks had eluded the gaze of outsiders for centuries. Yet the expedition, led by leading Italian explorer, the Duke of the Abruzzi, had unlocked the secrets of the Rwenzori, resulting not only in a series of first ascents, but also a detailed map. Perhaps the most abiding achievement of the mission however, had been the stunning set of cloud-free images produced by the greatest high-altitude photographer of the day.

Ever since the time of Ptolemy, in the second century CE, geographers had conjectured the existence of the so-called Mountains of the Moon in Central Africa, whose snowmelt was said to feed the waters of the Nile. Diogenes, a Greco-Roman merchant and traveller who landed unexpectedly in East Africa in 110 CE, had reported the existence of these mountains on his journeys in the region and ever since they had taken on mythical proportions, their imagined existence and inclusion on early maps of Africa being based on his sightings. After 25 days of moving along caravan routes in the African interior, Diogenes was said to have found two large lakes and a snow-capped range of mountains, which were claimed to be the sources of the great river. Historians suggest he was referring to Lake Victoria and another of the Great Lakes of East Africa.

Thanks to the work of Ptolemy, these supposed sources would become widely accepted as true by geographers for many centuries afterwards. As the contours of the continent started to take shape over the centuries, from the coastal regions

Nova Totius Africae Tabula: a map of Africa by Joachim Bormeester (d.1702) showing the assumed course of the Nile, the two great lakes (Zaire and Zafflan) and the Mountains of the Moon (Lunar Montes).

inwards, maps of the yet unexplored interior invariably included the *Montes Lunae*, and the two lakes, which bore the names of *Zaire* and *Zaflan*. When the heart of the continent was gradually opened up, this was all that early European explorers had to go on. Despite the fact that the existence of any such mountains must have been known to countless generations of locals who inhabited their foothills, their eventual 'discovery' went hand-in-hand with a fevered quest by Europeans for the sources of the Nile in the 19th century.

To the Victorians, the search to locate the source of the Nile was akin to sending a mission to the moon one hundred years later. Expeditions into the interior of Africa were fraught with unimaginable obstacles and dangers, the least of which was actually finding a route. Travellers could be struck down with a multitude of tropical diseases such as malaria, elephantiasis, hookworm, leprosy, yellow fever and sleeping sickness. Moreover, an East African slave trade was rampant in the second half of the 19th century, centring on the island of Zanzibar, from where caravans of heavily armed Arab slave traders would carry out raids on tribes in the interior, taking huge numbers of captives. Consequently, intruders of whatever origin, however benign, would be looked upon in fear and trepidation by ethnic Africans, and encounters often resulted in hostility and bloodshed. Neither was it uncommon for visitors to unwittingly get caught up in conflicts between warring tribes. Where European explorers were welcomed, they would be expected to pay homage to tribal chiefs, bearing opulent gifts. Often it meant being held up for weeks, even months.

There was no defining moment for the discovery of the Nile's sources, and though they have been mapped extensively, the river's uppermost headwaters are a subject of debate even today. Over a period of several decades from the 1850s onwards, Victorian explorers who embarked on this quest began slowly but surely filling in the missing pieces of the jigsaw. Conflicting theories as to the river's course abounded, and often the rivalry between explorers – and their armchair supporters back home – was so fierce, each of them presenting claims and counterclaims, that truth and fiction became intertwined. The simple fact was that before the various strands of river could be definitively joined up, it was a game of informed guesswork. More than anyone, it was Henry Morton Stanley (1841-1904) who most conclusively solved the riddle of the Nile's drainage system. Along with John Hanning Speke and David Livingstone, he is one of the three British-born explorers most closely associated with this quest.[1]

Between 1875 and 1876, after circumnavigating Lake Victoria[2], Stanley finally established that this vast tract of water had one single outlet, first 'discovered' by Speke in 1862 at Ripon Falls. On its long journey to the Mediterranean, the Nile flows into Lake Albert, some 200 kilometres downstream from Ripon Falls, another of East Africa's Great Lakes. Lake Albert itself is fed at its southern extremity by another substantial river system, the Semliki, which is likewise deemed to be a second major source of the Nile. To distinguish the two, they are often designated by the names Victoria Nile and Albert Nile.

It was only a matter of time before explorers would chance upon a region of snow and ice in these whereabouts: until the late 19th century, explorers travelling in the region had little inkling of the existence of a mountain range, enveloped as it was for much of the time in fog and cloud. In 1875 Stanley himself had camped in the eastern foothills of the range, but had been oblivious to its presence. However, on 24 May 1888, travelling south of Lake Albert, Stanley was alerted by a young boy to what seemed like the mirage of a mountain, capped by what he believed was a silvery cloud. On closer inspection he realised that he was in fact gazing at the form of a real mountain whose summit was covered in snow. He christened the mountain range the Rwenzori[3] as this was the closest transcription he could make of the name given by the indigenous peoples. It was Stanley's conviction that the mystery of Ptolemy's Mountains of the Moon had at last been solved, with their snow-capped peaks and glaciers feeding the headwaters of the Nile.

The Rwenzori mountains proved an enigma for European explorers in Stanley's wake. Between 1889 and 1906, a whole spate of expeditions had made attempts to unravel the mysteries of the mountain range. It was not until November 1905 that a serious mountaineering expedition was launched by Douglas Freshfield[4] and Arnold Mumm, but they too were defeated by the vagaries of the weather. The almost incessant cloud and the virtual impenetrability of a terrain overrun by luxuriant vegetation made it impossible to understand the rudimentaries of the topography, let alone find and climb any peaks.

Map showing the course of the Nile and its main tributaries in north-eastern Africa.

* * *

It is said that Luigi Amedeo di Savoia, Duke of the Abruzzi[5] (1873-1933), was first drawn to the Rwenzori when reading Stanley's obituary in 1904. It apparently made reference to a speech that Stanley had made to the Royal Geographical Society in London three years earlier urging *"some lover of Alpine climbing [to] take the Rwenzori in hand and make a thorough work of it"*. It was as if Stanley was appealing directly to the 31-year-old Duke.

The Duke, who was the nephew of the King of Italy, had developed a taste for travel and adventure at an early age and his status as a member of the royal household very much allowed him to indulge in these pastimes. In his youth he became a serious mountaineer learning his skills in the Italian Alps close to Turin, and later he circumnavigated the world as a naval officer in 1895. Exploration became his life's passion.

Two years later he organised a successful Italian expedition to Alaska, making the first ascent of Mount Saint Elias[6], at 5,489 metres the second-highest peak in both the United States and Canada. Based on these experiences in the Far North, he set out on a journey by ship to conquer the North Pole in 1899. Although the voyage failed to come home with the main prize, the expedition set a new record by reaching latitude 86° 34' north.

In 1905, one year after reading Stanley's obituary, and, having gained considerable acclaim and experience as an expedition leader, the Duke hatched a plan to explore the uncharted mountains of equatorial Africa. Although mountaineering was an intrinsic part of this latest plan, *"the main object of the expedition was geographical in the strict sense of the word; that is, to clear up the topography of the chain and ascertain the heights of the mountains. This work was to be completed by observations on physical geography, meteorology and magnetism."*

More than perhaps any other explorer of the day, the Duke was well placed to turn this into a successful venture. He was without equal when it came to meticulous detail and was astute in collecting the right people around him to help look after all the organisational, logistical and scientific aspects of an expedition. Moreover, as a member of the royal household, he could rely on the financial backing of the state, as well as wield considerable sway with the British: in the Scramble for Africa, the race by Western European powers to partition and open up the continent for their own interests, the British were colonising and developing much of East Africa and since 1894, Uganda - in which the Rwenzori were located - formed part of a British Protectorate.

The Duke intended to make his assault on the Rwenzori in May 1906, after the region's first rainy season. In the year that he had to put a campaign together, he assembled a team of twelve who, as well as contributing to the mountaineering element, would look after the scientific, that is, the topographical, photographic, botanical, zoological and geological aspects of the expedition. For the men's physical welfare there was a physician and a cook. In addition, he selected four alpine guides.

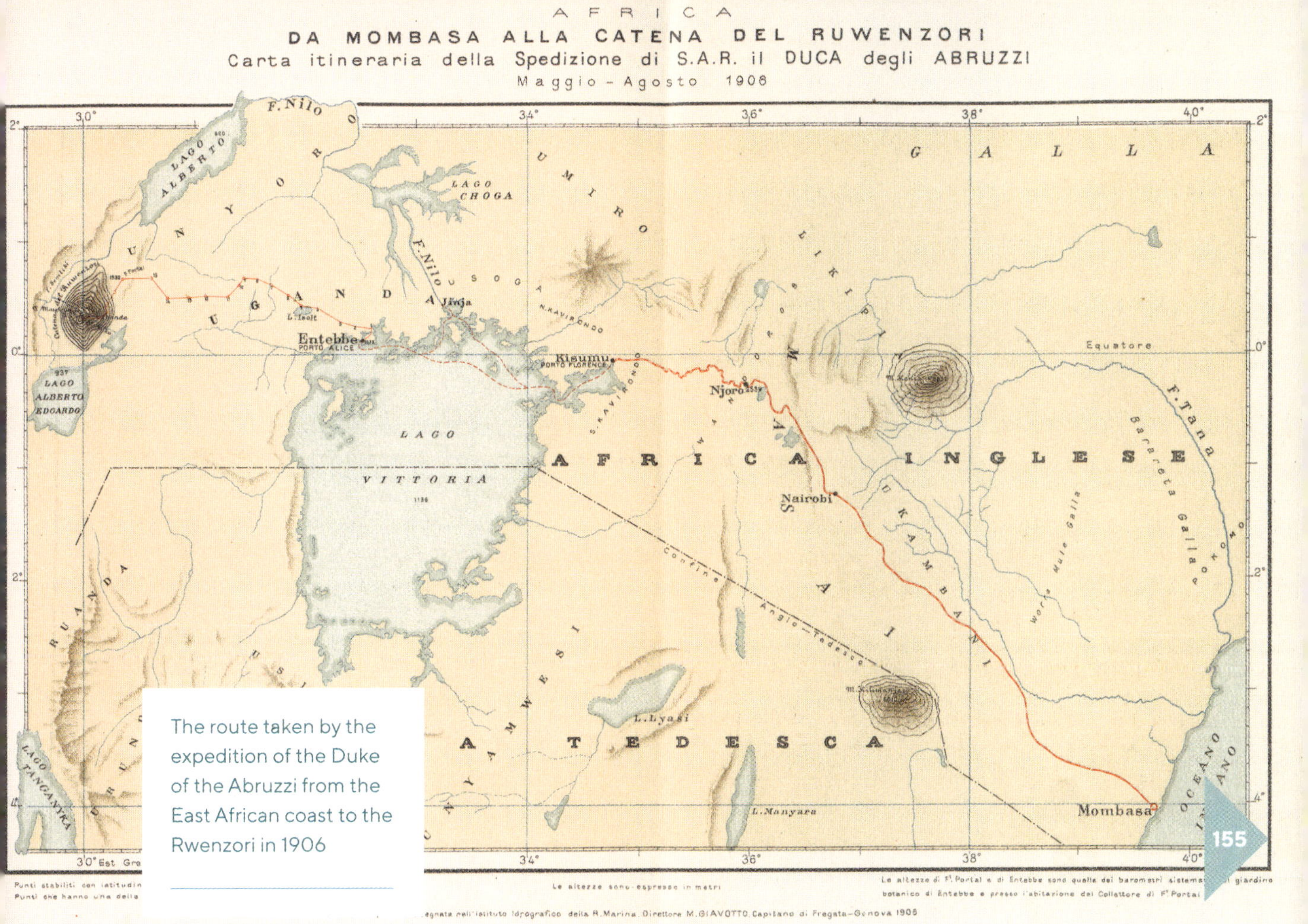

The route taken by the expedition of the Duke of the Abruzzi from the East African coast to the Rwenzori in 1906

Amongst the first to be invited to join the expedition party to the Rwenzori was Vittorio Sella (1859-1943) as photographer. Sella came from a well-to-do family of Piemontese industrialists who were also well-versed in the art of mountaineering.[7] Ever since a family acquaintance gave him a camera in his youth, Vittorio Sella had combined his passions of climbing and photography. He established a solid reputation as a mountain photographer and built up an impressive portfolio of images, not just from his own exploits in the Alps, but also from climbing outside Europe, including expeditions to the Caucasus, and the Duke's Alaskan campaign. By 1905, Sella was the Duke's go-to photographer.

Another indispensable member of the team was Umberto Cagni (1863-1932). Now aged 43, Cagni had acquired the status of admiral in the Italian navy and had served as the Duke's right-hand man on the Mount Saint Elias expedition in 1897 and the attempt on the North Pole in 1899. One of Cagni's roles would be to establish a baseline for the purposes of triangulation and to assist in the topographical work and the observations connected with physical geography.

On 16 April 1906, the party set sail from Naples for East Africa on the German liner *Bürgermeister*. The 6,600-kilometre voyage took them through the Suez Canal to the port of Mombasa where they arrived 18 days later. From there they took the metre-gauge Uganda railway[8], completed only five years previously, which would bring the team to Port Florence[9] on the eastern shores of Lake Victoria, over 900 kilometres distant on 6 May. Their onward journey then took them by steamer to Entebbe, at the time the administrative capital of Protectorate of Uganda, where they landed the next morning to be received by the British High Commissioner.

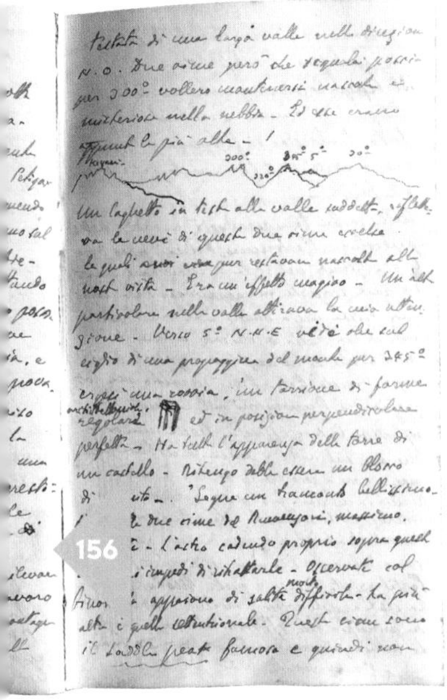

Notes from Sella's diary made on 11 May 1906, when the clouds suddenly cleared at Grauer's Rock enabling him to take his first panorama of the Rwenzori range

With no further means of motorised conveyance, reaching their final destination from Entebbe involved an overland footslog of over 300 kilometres and well-organised logistics for the paraphernalia of equipment, clothing and provisions needed for their mountain sojourn. To achieve all this, the Duke amassed a large troupe of local porters, mule drivers and servants numbering no fewer than 300. The caravan was ready to set off on 15 May, but now without Cagni, who had been stricken by malarial fever and lay in hospital in Entebbe.

It wasn't until 28 May, over 250 kilometres into their route march, that they first espied the snow-capped peaks of the Rwenzori, rising above the distant mists as if they *"were suspended in the air"*. On 7 June, 54 days after leaving Italy, they finally established a base camp at a place called Bugonjolo in the Mobuku valley, in the south of the mountain range. This base, the Duke believed, would give them the best springboard from which to climb the highest peaks. High above the valley at an elevation of 3,800 metres, the spot was described as a *"veritable eyrie"*, but it was in fact nothing more than a rock shelter whose constricted space had to accommodate the expedition team, the now vastly reduced number of porters, and all their equipment and provisions.

Like the few adventurers who had gone before them in this terrain, the Italians were soon to discover that exploring the Rwenzori was a stop-go routine. The mantle of almost incessant mist and cloud in which the landforms were swathed could render navigation impossible. Often, the patience of Job was required to wait for the vapours to dissipate and make the natural waymarks visible. The rain-laden skies also created an extraordinarily lush undergrowth, the likes of which expedition members had never encountered before. This description of the unworldly and untrodden landscape through which they had to toil their way is not untypical:

The bottom of the valley is nearly level, marshy, dotted with reeds, lobelias and senecio, and strewn with fallen trunks upon which you stumble at every step, and slippery with wet mosses in which you sink to the knee.

Camp in the Bujuku valley and Mount Gessi beyond.

Upper lake in the valley below the western flanks of Mount Baker.

The Duke of the Abruzzi (with cigarette) and Mr Knowles (British sub-commissioner of the Uganda province) after lunch during elephant hunting.

Vittorio Sella with Mr Knowles' dog near Moore Glacier.

However, the Duke was not one for dallying, and two days after setting up base-camp he was heading into the mountains. On 9 May, the first morning of climbing, he attained a height of 4,349 metres, high above the Mobuku Glacier at the head of the valley. Here, his party was halted by a cold, damp fog which persisted for the rest of the day, and he was forced to pitch camp. Before daylight the next morning, with clear skies and an *"irresistible impatience to proceed"*, the Duke then ascended to the head of the glacier before reaching a col at the base of a rocky crag on a summit ridge dubbed Grauer's Rock.[10] This clear weather at the start of the expedition was a magnificent stroke of luck, since the uninterrupted view from this vantage point was decisive in unlocking the mysteries of the Rwenzori's topography.

It was the first time that any expedition had been able to appreciate the full extent of the mountain range, as previous travels had been hampered by its swirling mists. From this point on the ridge line of the Mount Baker group, on which the Duke was now standing, it was possible to discern five other distinct mountain groups in the distance. The nearest of these groups, to the north west – now referred to as Mount Stanley - was entirely surrounded by snow peaks and glaciers, and was determined to be the loftiest of the Rwenzori.

That same day, further exploration of the western ridge of Mount Baker and the first ever ascent of its highest point, Edward Peak (4,844 metres), also revealed a route by which Mount Stanley could best be approached. The next day, 11 May, the Duke and his party returned to Bugonjolo, leaving Sella and his helpers to go up to the col to take photographs.

Sella had worse luck and spent several hours on the col waiting for the mist to clear. He climbed the adjacent rocky crag, where he set up a photographic camera on its tripod. Then, the party huddled around, waiting patiently. By two in the afternoon Sella gave up, folded up the camera, and was on the point of leaving, when suddenly the heavens began to clear. Within the space of a few minutes all the mountains were uncovered and the camera was immediately set up again to take a panorama.

Over the next few days, as the official account recorded, the expedition was confined to camp owing to the atrocious weather:

The rain was nearly continuous, while storms of wind, with thunder and lightning, followed upon one another at short intervals. Thick dark fog enveloped everything. The camp was soon invaded by mud and water, and a continual drip came down upon the tents from the overhanging rock. Every time that anyone stepped out of the camp he would sink into the mud.[11]

On the morning of 15 May, the skies cleared and, without delay, the Duke assembled a team of guides and porters and set off to follow the line of approach to Mount Stanley he had seen a week earlier. Crossing the watershed at Freshfield Pass, the party encamped close to one of two lakes that nestled in the adjoining valley on the western flanks of Mount Baker. They spent the next two days ferrying provisions before establishing a high camp at 4,516 metres on the ridge overlooking the Elena Glacier on Mount Stanley.

Cima Alessandra (left) and Cima Margherita of Mount Stanley from the eastern ridge of Cima Alessandra.

The next day they broke camp before daylight and by 6.30 am they reached the summit plateau, which gave clear views of the two summit peaks at the northern end of the group. The nearest, which they discovered was the lower of the two, they were able to ascend with relative ease, but then, bedevilled by fog, they had to wait to get a clear sight of the way ahead to the higher peak. Eventually they decided to take the shortest and most direct route (in retrospect, almost certainly the most difficult). Descending a steep slope to the saddle between the two peaks, they then crossed a crevasse-riven glacier before having to cut through a cornice to gain the summit. It was not yet midday and the Duke and his three companions, guides César Ollier and Joseph Petigax[12] and porter Joseph Brocherel had conquered the Rwenzori. The Duke unfurled the Italian *tricolore* and planted it in the snow on the summit, 5,109 metres above sea level. He christened the highest summit *Cima Margherita* in honour of Margherita[13], former Queen and now Queen Mother of Italy. The lower summit (5,091 metres) he named *Cima Alexandra* after the Queen Consort and wife of King Edward VII of Great Britain.

Meanwhile, after being discharged from hospital in Entebbe on 5 June, Umberto Cagni had arrived posthaste to join the expedition in Bugonjolo on 16 June. Cagni was integral to completing a map of the Rwenzori, one of the chief objectives of the mission. There were two elements to this: first, a triangulation of the mountain range which connected its numerous peaks to a baseline; and second, the determination of the exact latitude and longitude of the Rwenzori through astronomical observations to fix its position on the map of Africa. Evidently, all this depended on perfect ground conditions and clear skies.

Delineating a 300-metre baseline, which was deemed sufficient to give meaningful differentiations in angles and affording clear views of two nearby peaks – Edward Peak and Cagni Peak[14] – from which further angles would be taken, was clearly going to present a problem. Finding a level stretch of ground was hampered by mud, carpets of moss and the giant senecios dripping with rainwater. Ultimately, at each extremity of the baseline, they had to make piles from tree trunks sunk into the ground to provide solid foundations for a theodolite. As for celestial observations, at the end of June, Cagni was plagued by a complete absence of sun and it was only during rare moments of breaks in cloud cover that he was able to make a small number of astronomical measurements.

Remarkably, over the course of the next few weeks, the Duke and his companions continued to make first ascent after first ascent of all the major peaks. This was, of course, not the sole objective, but in mountaineering terms it was the most prestigious achievement of the expedition. Many of the peaks were climbed multiple times[15] so that the secrets of the Rwenzori's topography could be more definitively uncovered.

In this respect, the expedition member with the greatest patience was Vittorio Sella: he would often find himself re-ascending peaks and cols to attain vantage points which would offer the perfect shot without any intervening cloud. According to the official account of the expedition, during one week at the beginning of July, Sella camped at Freshfield Pass (4,326 metres) making *"fruitless expeditions up the ridges, [with] hours and hours of waiting beside his camera in the storms. For all his tenacity and energy he had not succeeded in getting a complete panorama from the Edward Peak as he had proposed to do."*

But the Duke too, worked tirelessly to help unlock the puzzle of the Rwenzori. Whilst Cagni made use of a heavier theodolite, the Duke was busy taking compass readings from many of the peaks of the range to measure their angles relative to each other, and these measurements were often repeated over and over again for the purposes of verification.

By 16 July, they had climbed all the main peaks in the Rwenzori and their time in the mountains was coming to an end. It was another month before they had retraced their steps back to Mombasa on the coast.

On their return to Italy, the Duke and his men were given a rapturous welcome. After his previous successes in the Far North, Luigi Amadeo's triumph in Africa had increased his standing at home to new levels. He was the talk of the Italian newspapers for weeks and his fame now spread far beyond the confines of the Italian peninsula. He became a much desired speaker and in January recounted his experiences to packed-out audiences at the *Società Geografica Italiana* in Rome and the Royal Geographical Society in London.

The survey confirmed not only that the Rwenzori lie less than half a degree north of the equator, but also that the whole range sheds its waters into the Nile river basin, whether this be via the Victoria Nile on its eastern flanks or the Albert Nile on its western. Was this the final proof that the Rwenzori were indeed the mythical Mountains of the Moon? With a general north-south orientation, the map

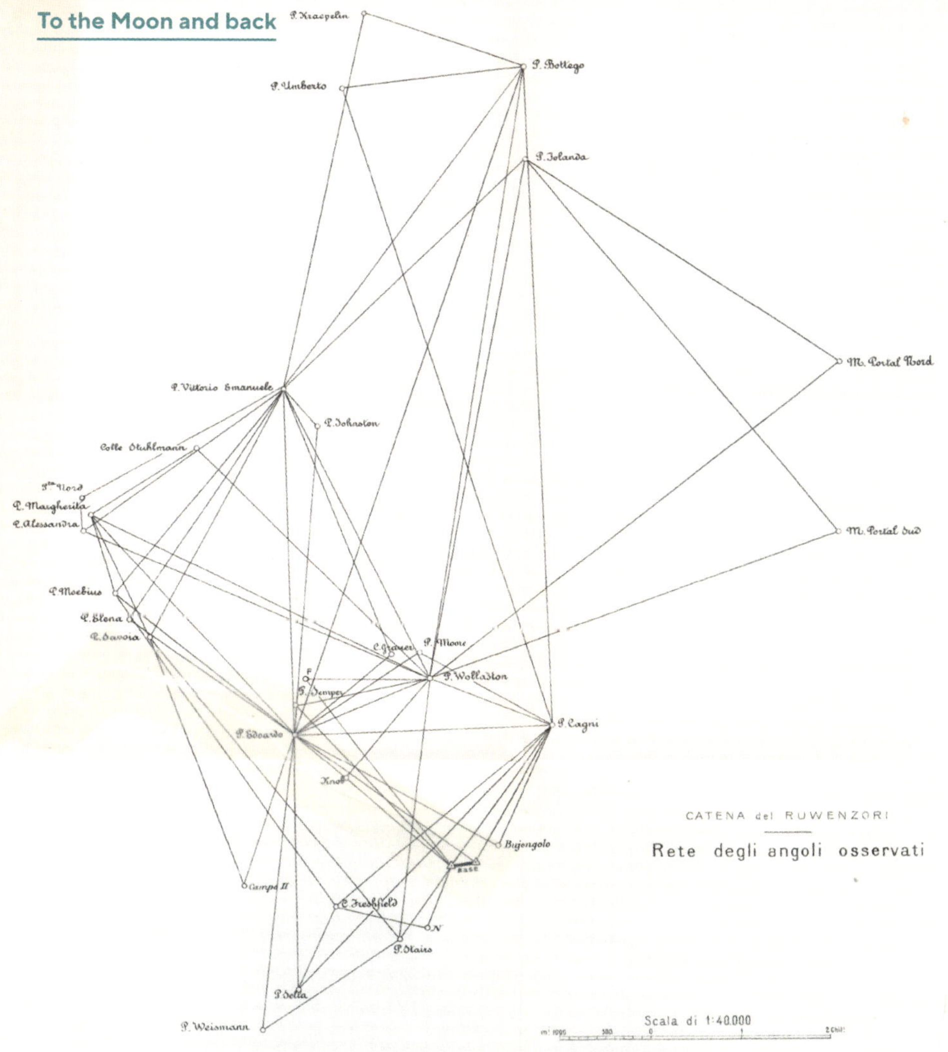

Cagni's triangulation of the Rwenzori (with baseline in the south).

also shows that the Rwenzori range is made up of six distinct mountain groups of peaks and glaciers divided from each other by snow-free cols. It was the map that was deemed to be the most important and successful geographical outcome of the expedition.

The most enduring legacy though must surely be the stunning photographs that Vittorio Sella has left us with. He managed to capture on film all the mountains of the Rwenzori (most of which he also climbed) from different perspectives. Even more remarkable is the fact that he managed to find the right views and vantage points in a hitherto unexplored region under a thick and almost constant sheet of cloud. Looking at his photos, you can almost sense the thrill and excitement he must have felt when – after days of gloom – the clouds parted to reveal a backdrop of equatorial snows. However exceptional his mountain photos may be, his work in the Rwenzori should not just be remembered solely for these: Sella also made botanical studies in close-up and documented the local people. He returned to Italy with a timeless snapshot of life in the Mountains of the Moon at the turn of the 20th century, the likes of which had never been witnessed before.

* * *

An expedition porter.

Notes on the nomenclature of the Rwenzori

According to the Konjo people who inhabit the Rwenzori, the peaks are considered a sacred personification of their guardian spirits. The supreme deity, *Kitasamba*, is said to live on the glacier-covered summits and the ice represents the frozen sperm of the god, which melts into the rivers and streams to fertilise the land. The 1906 expedition broke a taboo that prevented humans from reaching the ice, so their presence and the nomenclature for the peaks symbolically appropriated the Konjo's natural sanctuaries.

'Rwenzori' - sometimes written 'Ruwenzori' - was the name that Stanley believed most closely approximated the word that the locals called the mountains. His own name was later given to the highest massif in the Rwenzori, as Mount Stanley. Local names already existed for some of the other mountain groups, for example, Kiyanga and Duwoni, which were re-christened Mount Baker and Mount Speke (named after two of the leading Victorian explorers).

When the Duke of the Abruzzi's expedition was tasked with naming the various topographical features, they found it difficult to trace any local names for individual peaks: only a few names of rivers and individual places existed, such as the Bujuku and Mobuku rivers and Bugonjolo, in the valleys inhabited by the Bakonjo tribe on the eastern side of the watershed. Consequently most of the names given to the peaks, passes and glaciers were designated by the Duke and refer either to the names of those who had previously explored the region (such as Freshfield Pass, Scott Elliot Pass, Mount Gessi, Stairs Peak and Grauer's Rock), to commemorate royalty (Margherita Peak, Alexandra Peak, Umberto Peak and Edward Peak), or were named after expedition members themselves (Luigi Amadeo, Sella, Cagni and others).

It should be added that, at the time, the names for all the geographical features had to be approved by the British authorities. Unlike some mountain regions which were colonised by Europeans, modern maps of the Rwenzori overwhelmingly retain the names from this period.

Endnotes

1 John Hanning Speke (1827–1864) was the first to produce compelling evidence that Lake Victoria fed the Nile when he discovered the great outflow at Ripon Falls on its northern perimeter in 1862 after tracking its eastern shores. But for the fact that he fatally shot himself in a hunting accident back home in England, Speke would have probably gone on to uncover the true course of the Nile much sooner. David Livingstone (1813-1873), Britain's most intrepid explorer of the Victorian age who had travelled widely through the African interior, was of the opinion that the Nile followed a much more westerly line rising in the vast reservoir of Lake Tanganyika, flowing out at its most northerly point and continuing northwards along the Rift Valley via Lakes Kivu, Edward and Albert. This theory was quashed conclusively in 1871 when, sailing to the northern extremity of the lake, Livingstone and Henry Morton Stanley discovered that there was no outlet. Undissuaded, Livingstone, who had made it his life's mission to uncover the course of the Nile, conjectured that the outflow on the western shores of the lake might feed its upper waters, but died before this theory could be disproven. It was Stanley on his trans-Africa expedition of 1874 to 1877, who discovered that Lake Tanganyika was in fact one of the sources of the Congo river.

2 Named at the time after the British queen, Lake Victoria is so large that it has multiple local language names, such *Nam Lolwe, Nnalubaale, Nyanza and Ukerewe*.

3 This local name purportedly means "Rain-Maker" or "Cloud-King". Given the fact that it rains for an average of 350 days a year, the name is hardly unbefitting.

4 Douglas Freshfield (1845-1934) was a British mountaineer and traveller. He was accompanied by photographer Vittorio Sella on expeditions to the Caucasus in 1896 and Kangchenjunga in 1899.

5 His full Italian title at the time was 'Sua Altezza Reale, il Principe Luigi Amedeo di Savoia, Duca degli Abruzzi'.

6 In Tligit, the native language, the mountain is known as *Yas'éit'aa Shaa*, or 'mountain behind the Icy Bay'.

7 Sella's uncle Quintino Sella, apart from being one of the foremost politicians in post-Risorgimento Italy, was co-founder of the *Club Alpino Italiano* (CAI).

8 Built at enormous cost to connect the heart of British East Africa with the Indian Ocean, the line was nicknamed – ironically in the context of this story - the 'Lunatic Express', on account of the large numbers of fatalities amongst construction workers who succumbed to disease, tribal hostilities, and most notably to man-eating lions. One estimation puts the number of those mauled to death in this way at 135.

9 Now Kisumu, in present-day Kenya.

10 Named after Austrian mountaineer Rudolf Grauer, who had reached this point in January 1906.

11 It was around this time that an unwelcome intruder was spotted prowling the margins of the camp: a large leopard was observed devouring two sheep and on another occasion almost entered the Duke's tent before being frightened away. The story didn't end well for the leopard. Two members of the party planned to lure the animal into a trap entailing a piece of meat and two rifles and one night it was killed on the spot with a bullet through its skull.

12 Joseph (or Giuseppe) Petigax (1860-1926) was a renowned guide from Courmayeur in the Val d'Aosta who served with the Duke of the Abruzzi on several expeditions, including the Italian expedition to K2 in 1909 achieving a new altitude record of 7,498 metres.

13 Queen Margherita (apart from being most commonly associated with the eponymous pizza) was also a keen mountaineer and became honorary president of the Ladies' Alpine Club. Founded in London in 1907, it was the first mountaineering club for women.

14 Named during the expedition in honour of Umberto Cagni.

15 Alexandra Peak was climbed no fewer than five times during the expedition.

Swiss-style
Snapshots from a nation's cartographic past

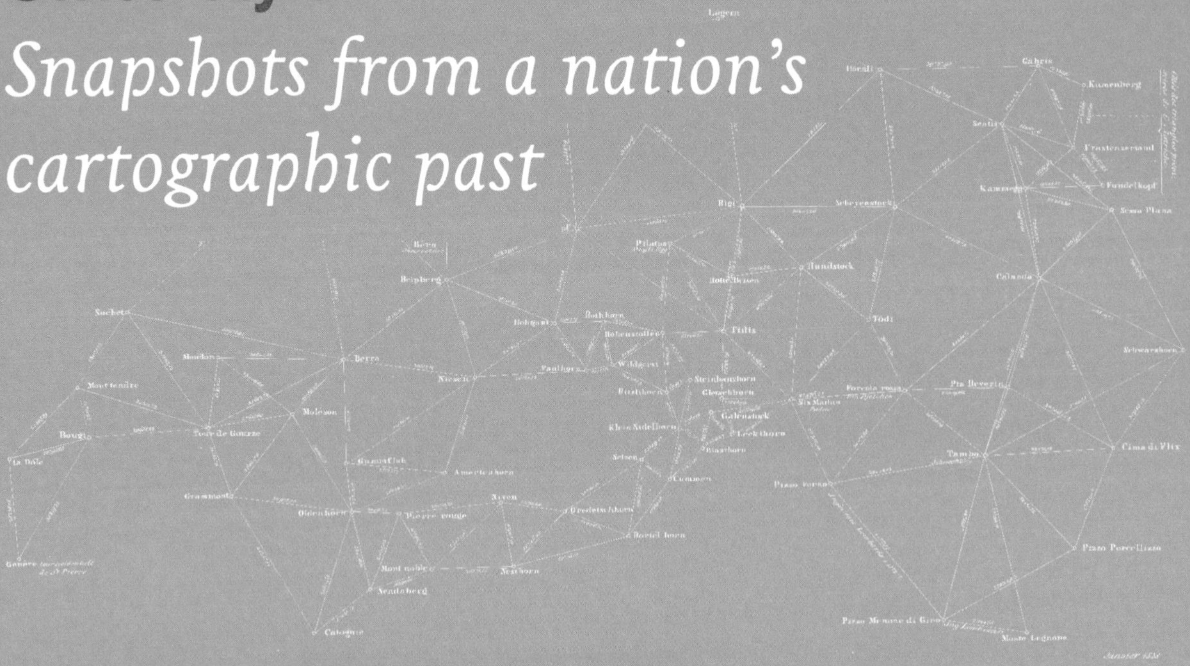

Primordial triangulation of Switzerland (1834/1837).

After the fall of the ancien régime in France and the turbulent decades that followed, as with many newly emerging states across the continent of Europe, Switzerland too found itself on a course to a more unified federal state and there was a growing need to map the country, not least for military ends. One man in particular was responsible for drawing up a cartographic plan – Guillaume Henri Dufour.

There are perhaps few nobler accolades that could be bestowed on a cartographer than having the highest point in the country named after you. The *Dufourspitze*, the highest peak in the Monte Rosa range in Switzerland, is one such peak.[1] Born in Konstanz but having spent his formative years in Geneva, Guillaume Henri Dufour (1787–1875) studied at the *École Polytechnique*, the French military academy in Paris, and in 1811 joined the French army. In 1817, he returned to Switzerland as commander of the Canton of Geneva's military engineers and a teacher of mathematics at the city's academy.[2] In 1832 he was appointed Quartermaster-General of the Swiss Army, whereupon his duties also included nationwide surveying. He was entrusted with the task of creating a topographic map of the whole country at a scale of 1:100,000. Working with astronomer and geodesist Johannes Eschmann, between 1834 and 1837, a triangulation network known as the *Triangulation primordiale* was established which for the first time joined areas to the north of the Alps with those to the south. This trigonometric survey would serve as the reference system for the so-called Dufour Map. Alongside this triangulation, topographic surveys were also initiated which gathered together survey data from the different cantons in Switzerland so that the information could be used to fill in the triangles and then converted to the scale of the envisaged map. Consisting of 25 sheets at a scale of 1:100,000, the Dufour map of Switzerland was created between 1845 and 1865. In view of its mountainous terrain, the map is especially significant for the use of a technique known as oblique-light hachuring to depict the relief. This was a form of shading that gave the landscape a distinct three-dimensional impression, particularly suited to regions of strong relief and sharply defined ridges,

Sheet 22 (Martigny-Aoste) of the 1:100,000 Dufour map of Switzerland published in 1861.

such as in Switzerland. This method of hachuring raised the benchmark for mountain mapping to even higher levels and won Dufour's topographic bureau several international awards. Even today, the Dufour map is venerated as one of the most important symbols of Swiss unity and hangs in the visitors' entrance of the Swiss parliament building in Bern.

One of the most colourful figures in Swiss cartography is surely Xaver Imfeld (1853-1909). Imfeld was born in Sarnen, a small town 20 kilometres south of Lucerne, to where his family later moved when he was fourteen. He studied surveying at the Federal Institute of Technology (ETH) in Zürich and by the age of 23 had joined the *Eidgenössisches Topographisches Bureau*[3], the federal mapping agency in Bern. This was during a period when the 1:100,000 Dufour map of Switzerland was being superseded by the Siegfried map[4] whose intention was to cover the country at the larger scales of 1:50,000 in the mountain areas and 1:25,000 in the rest of the country. This was primarily to meet the requirements of the military, as well as industry, but also the growing demands of the mountaineering fraternity, embodied by the Swiss Alpine Club (SAC, founded in 1863).

Employed on a contractual basis, Imfeld was set to task working on 21 sheets in Central Switzerland, the Bernese Oberland and Valais, which led to him specialising heavily in mountain topography, particularly the depiction of rock features. His summers would consist of long hours spent in the field whilst in the winter his work was confined to the drawing office. Though he excelled at cartographic skills, a career in surveying did not pay well and Imfeld – partly due to the fact that he had seven children[5] – was compelled to take on additional work to supplement his income, in the first instance by creating relief models and panoramas. As early as 1877/78, for example, he had produced a 1:25,000 relief model of the Zermatt Alps and a sublime panorama of the Monte Rosa group.

From 1888, he became a freelance surveyor and branched out producing tourist maps, panoramas and relief models, but he also became involved in road and railway engineering projects. For example, in 1891 he spent three weeks working with Gustave Eiffel on a project to build a permanent observatory on the top of Mont Blanc, which failed to materialise partly because the mass of snow and ice on the summit prevented them from identifying any foundation anchors.[6] Nevertheless, his experiences on the mountain eventually resulted in a spectacular four-sheet panorama from its summit in 1895, and a masterly relief map of the Mont Blanc massif at a scale of 1:50,000 in 1896. Alongside, Pfyffer's *Urschweiz*, the *Gletschergarten* museum in Lucerne also features relief models by Imfeld which it commissioned.

He also worked on the *Jungfraubahn*, the highest rack-and-pinion railway in Europe, built between 1896 and 1912[7], but sadly never saw its completion. Probably due to the lasting effects of the demands placed on his body from his experiences on Mont Blanc from which he never fully recovered, Imfeld died of a heart attack in 1909 at the age of 56. During his lifetime, he received numerous commendations for his work from home and abroad and, for his accomplishments as mountaineer and surveyor, was granted honorary memberships of the Swiss, German-Austrian and French alpine clubs.

360-degree panorama of the Monte Rosa Massif taken from the Dufourspitze by Xaver Imfeldf in September 1878.

Sheet 2 of a four-sheet series of the 1:50,000 map of the Mont Blanc Massif drawn by Xaver Imfeld and published in the 1906 journal of the Swiss Alpine Club.

Endnotes

1 At 4,634 metres, the *Dufourspitze* straddles the Swiss-Italian border and is the second highest peak in the Alps by prominence. Maps at the time indicated the peak as *Höchste Spitze* before it was decided to rename the mountain in honour of Dufour in 1863 following completion of his map.

2 Dufour was appointed to the position of General in 1847 and that same year helped subdue an attempt by several cantons to prevent a more unified federal state (known as the *Sonderbund* war). In 1863 he sat on a committee which ultimately led to the foundation of the International Red Cross and also presided over an international conference which helped bring about the 1864 Geneva Convention.

3 The forerunner of today's *swisstopo*, the national mapping agency

4 The map was named after Hermann Siegfried who succeeded Dufour as chief of staff at the federal mapping agency. Publication of the Siegfried editions commenced in 1870, but were not completed until 1922 (1:25,000, 462 sheets) and 1926 (1:50,000, 142 sheets) respectively.

5 Imfeld married Marie Seiler, daughter of hotel pioneer Alexander Seiler who operated the famous Hotel Monte Rosa in Zermatt where Edward Whymper often stayed during his attempts to scale the Matterhorn.

6 The expedition members were based in the original Cabane Vallot referred to elsewhere in this publication (see Helbronner). One of their company, Doctor Jacottet, died of respiratory failure which was the first documented case of pulmonary oedema resulting from prolonged exposure to high altitude.

7 Following the first ascent of the Matterhorn in 1865, the fame (and infamy) of the mountain helped bring tourists in their thousands to Zermatt and in 1890 Imfeld even came up with a plan to construct a railway and cable car to the summit of the peak. Needless to say, the project never saw the light of day, but construction of the mountain rack-and-pinion railway from Zermatt to the Gornergrat was completed in 1898.

Published by Edward Arnold & Co for th
from the maps prepared by the Roy

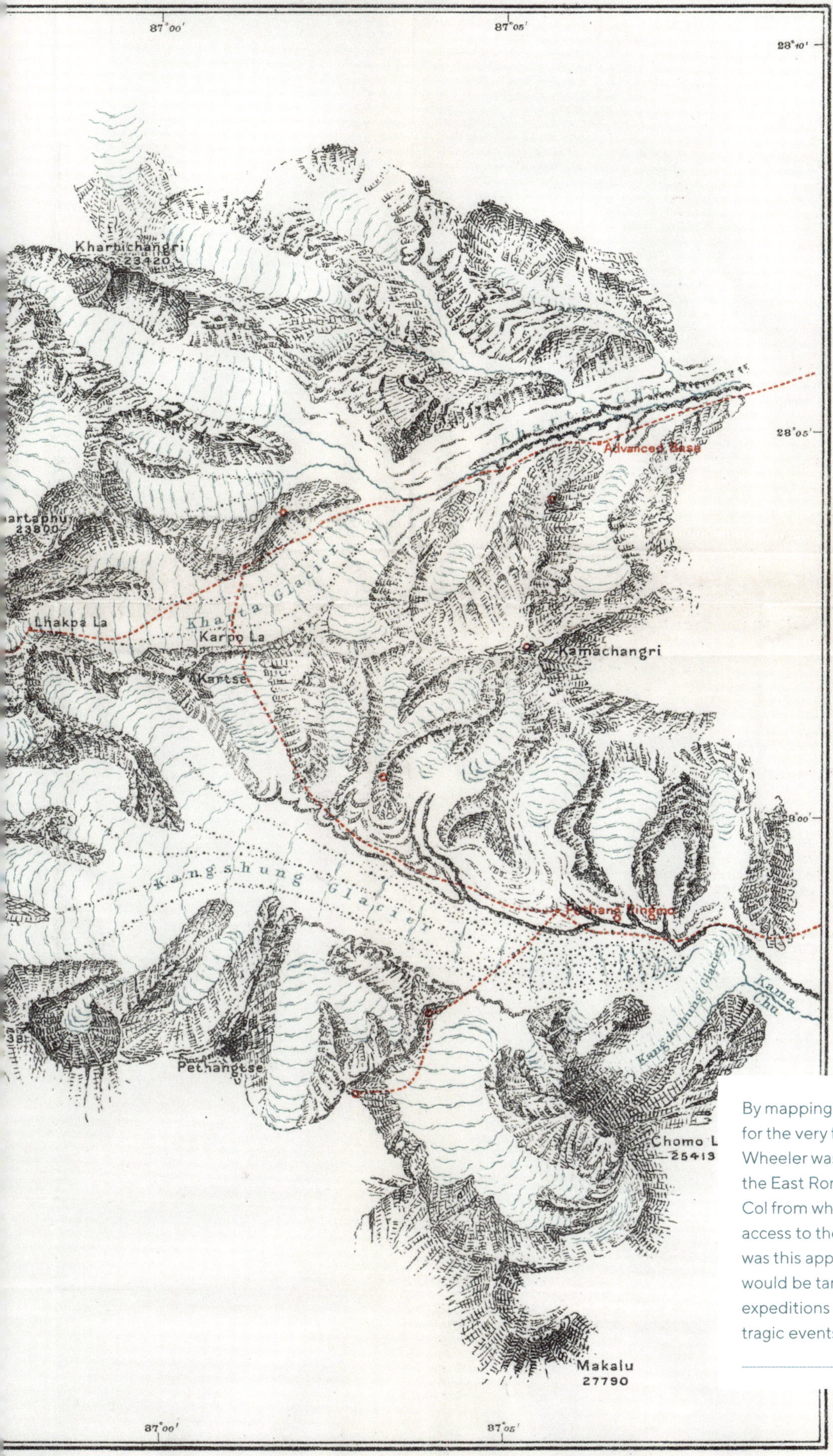

By mapping Everest at close quarters for the very first time, in 1921 Oliver Wheeler was able to identify a route via the East Rongbuk Glacier to the North Col from where it was possible to gain access to the north ridge of Everest. It was this approach to the summit which would be targeted on subsequent expeditions and ultimately result in the tragic events on the mountain in 1924.

Chapter 15

Everest finally on the map

The first ever reconnaissance mission to Everest in 1921 surveyed the mountain from close-up and revealed a possible approach to its summit

The mountain itself has spawned a global industry: thousands of tourists flock to its base camp each year, and hundreds more - with bigger egos and deeper pockets - aspire to reach its summit. Yet barely a century ago, Mount Everest was described as *"shy and retiring, hidden behind other mountains"*.[1] Situated on the border between two of the most secluded and closely guarded countries on the planet, with the exception of a few intruders who came back with a few rough notes and sketches, the environs of the world's highest mountain were cloaked in mystery.
In the ensuing quest to find a way to its summit, a plan was put in place to lay bare its secrets.

It wasn't until 1921 that the first ever detailed map of Mount Everest was made. 65 years previously, the mountain - then known as Peak XV - had been conclusively identified as the world's highest by the British Survey of India, the result of observations taken from six stations in the distant Indian plains. In 1856, this appellation was replaced by 'Mount Everest' on the instructions of the Surveyor General of India, Andrew Waugh, in honour of his predecessor Sir George Everest.[2] Measured at 29,002 feet (or 8,840 metres), it marked the culminating point, both literally and metaphorically, of the Great Trigonometric Survey which had mapped much of the Indian sub-continent from south to north.

After this proclamation, Mount Everest, straddling the border between Nepal and Tibet, remained unexplored at close range for six more decades. This was because Nepal remained out of bounds to all foreigners, a ban that would remain in place until 1949. On the other side, to enter Tibet - for centuries a forbidden land - diplomatic permission was required. By the end of the First World War however,

Britain was keen to prioritise the conquest of Everest. Having failed to win the race to both the North and South Poles, the world's highest mountain was looked upon as the 'Third Pole'.

In January 1921, a Mount Everest Committee was formed jointly by the Royal Geographical Society (RGS) and the Alpine Club, presided over by Sir Francis Younghusband.[3] At the time, no explorer had got to within a hundred kilometres of Mount Everest, so it was imperative that the mountain could be observed from close-up. The prime objective of the committee of course was to put a British team of mountaineers on the summit, but given their ignorance of the terrain, even they realised that this wouldn't be possible in one season. It was first necessary to get a detailed understanding of the mountain's topography. To all intents and purposes, there were two sides to the mission - mapping and mountaineering - which happened to coincide with the interests of the RGS and the Alpine Club respectively. It was decided a first expedition would focus efforts on surveying the mountain and, on the basis of the results, a plan for an all-out assault on the mountain could be made during a second season.

After much diplomatic toing-and-froing, the British were eventually granted permission by the Dalai Lama to enter Tibet, and in 1921 a military-style reconnaissance mission was mounted under the leadership of Colonel Charles Howard-Bury. His primary objective was to explore the northern approaches to the mountain with the aim of finding a route to the summit. The lead climber on the 1921 expedition team was George Mallory.

The remit for the survey part of the 1921 reconnaissance mission was as follows:

Mount Everest is flanked to the west by a high group [of mountains] reaching to 26,750 ft, to the north by a group reaching 23,640 ft, to the east by the Makalu group (27,850 ft) and to the south by many high peaks. Permission to enter Nepal not being available, only the area north of the boundary could be considered. A map which will explain Mount Everest clearly and will show its connection with the neighbouring mountains and with the drainage of its vicinity, must include the mountain groups mentioned above. In fact they, with Mount Everest itself form one great mountain mass, the whole of which may be described as the Everest group.[4]

The person chosen for this cartographic survey was the Canadian, E.O. (Oliver) Wheeler. Under the command of his senior Henry Morshead, he was assigned as secondary surveyor to the mission but was otherwise charged with the detailed mapping of Everest at a scale of one inch to one mile (1:63,000). Having served with the British and Commonwealth forces in the First World War, in the years that followed Wheeler found himself working at the headquarters of the Survey of India, based in Dehra Dun[5], so was geographically well-placed to fill this position. But not just geographically: he came from an illustrious line of surveyors and alpinists who had helped open up the Canadian Rockies with the aid of photogrammetry, at the time a revolutionary new surveying method. Wheeler described this 'Canadian' method as "*plane tabling by photography*", the camera replacing the field-surveyor's eye and cartographic details being filled in on the basis of the

Wheeler (left) and his assistants taking a break.

photographs. This technique lent itself in particular to high-altitude mapping, for example: much larger areas could be covered; field data could be transferred onto maps afterwards in the office; and, not unimportantly, *"there is no necessity to do accurate drawing [in the field] with numbed fingers"*.

Numbed fingers or not, surveying over 3,100 square kilometres of country in some of the world's most inhospitable terrain was not going to be a walk in the park. Starting out on 18 May 1921, the month-long trek from Darjeeling, through Tibet to the base of Everest at the snout of the Rongbuk Glacier, was 480 kilometres. The rest of the five-and-a-half-month campaign was going to present a huge logistical challenge for the party, made no easier when they lost one of its key members, Alexander Kellas, to heart failure on 5 June. More than anyone on the expedition, Kellas had an intimate knowledge of climbing at high altitude in the Himalaya.[6] It was a cruel blow.

For his part in the reconnaissance mission, Wheeler required heavy surveying equipment (camera, theodolite, tripod, levelling equipment and fragile glass negative plates) which amounted to almost 50 kilos in weight, along with tents, food and other provisions, all transported to isolated stations, often above 6,500 metres.

Once they had reached their destination, work could begin. Wheeler operated separately from the rest of the reconnaissance party, often spending days on end waiting for the weather to clear before he could take readings and photographs. He was supported by three high-altitude assistants, who shared a tent, and ten permanently engaged porters, who also acted as mail runners and fuel gatherers. None of the porters were fitted out with warm clothes or tents and they had to sleep at the main camps, ferrying stores up to Wheeler's survey stations or moving camp.

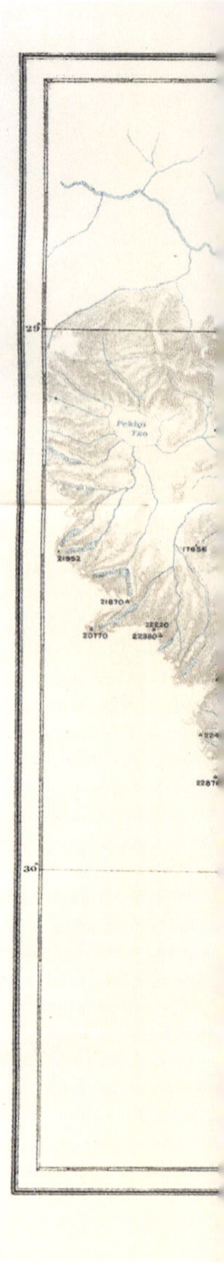

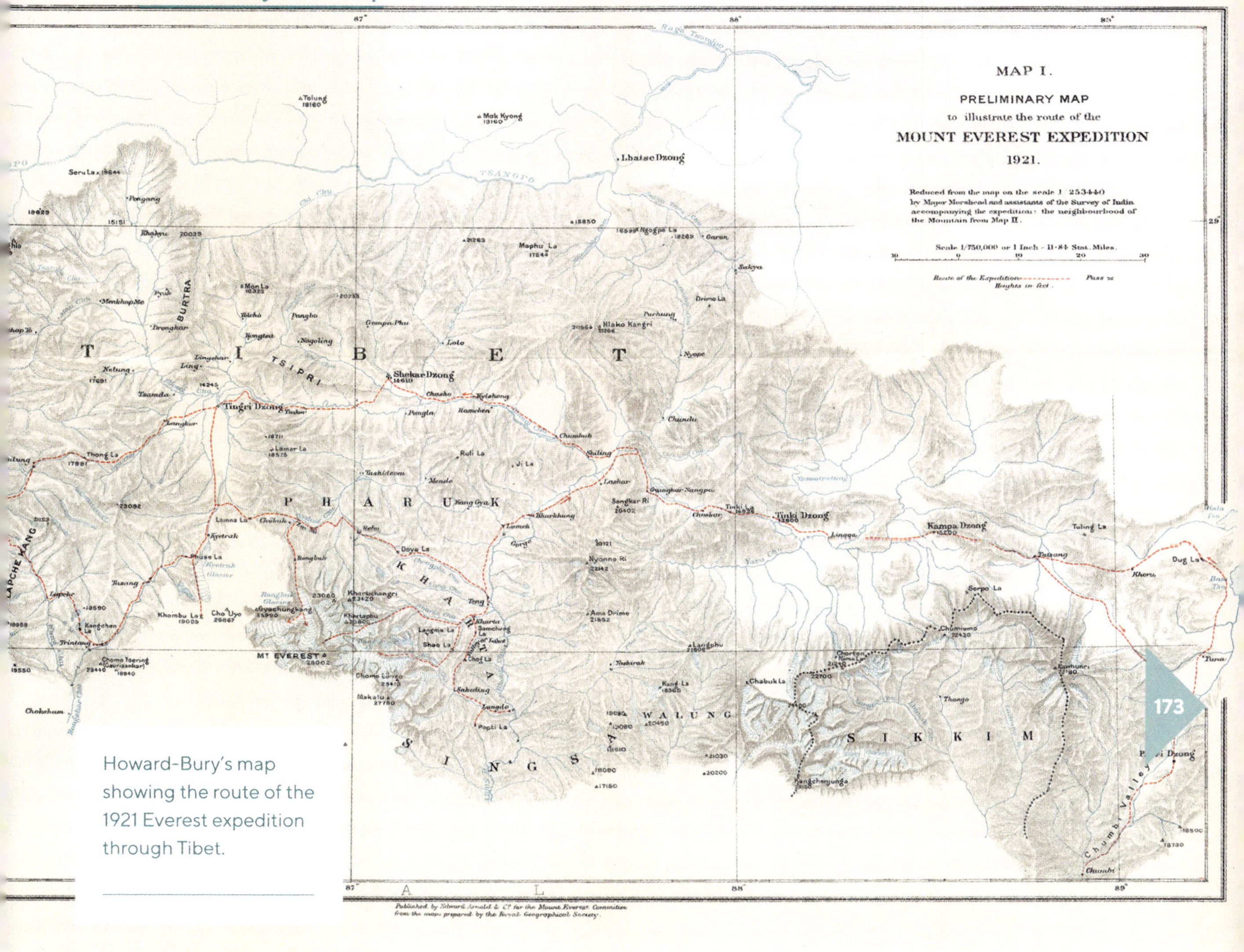

Howard-Bury's map showing the route of the 1921 Everest expedition through Tibet.

Often Wheeler would camp out on the mountain alone. By the time the British reconnaissance team set out on their return to Darjeeling in late September, Wheeler had spent 41 nights on moraines and glaciers at altitudes between 18,000 and 22,200 feet. He shot 240 images, which would sometimes be developed in tents on dark nights at high altitude, or during occasional breaks at base camp where he would once again be able to enjoy the company of fellow expedition members.

In one account not untypical of his sojourn in the high mountains, Wheeler recounted how he arrived back into a camp on the West Rongbuk Glacier one day soaked to the skin. He wrote:

I was ... [here] ...for five days: most of them spent huddled under rocks waiting for the clouds to lift. I had one beautiful day, my only one in six weeks, and got some very nice photographs of Mount Everest and its West ridge. It is surprising how a little good weather and the feeling of having really done some work affects one's spirits!

Wheeler was ultimately credited with unlocking a route to the North Col via the East Rongbuk Glacier, which other expedition members (like the begrudging Mallory) had failed to identify in their reconnaissance. From the North Col, it was possible to access the north ridge of Everest and it was this approach to the summit which would be targeted on subsequent expeditions. Towards the end of the expedition in late September, Wheeler himself attained a height of 23,000 feet (or 7,000 metres) climbing with a group headed by Mallory. In fact, were it not for the sapped energy of fellow team members, Mallory would have pressed on in 1921. Eventually, it would be left to the expeditions of 1922 and 1924 to try and find a way via the North Col and the North Ridge to the summit, the latter expedition ending in the disappearance of Mallory and his climbing companion Sandy Irvine.

On a personal level, Wheeler found much of his own expedition monotonous and lonely, but relished the few times when there were clear days. Overall, he said, "*I would have not missed the 'show' for anything.*" In his survey report he concluded, "*On the whole we felt that we were able to pass on to the 1922 expedition*[7] *as complete information as it was possible to get in the time available, and to give them maps, photographs and general information sufficient to enable them to make a detailed plan of campaign for the final assault.*"

Morshead was not quite so reserved in assessing Wheeler's achievements however, writing in the official account of the 1921 Mount Everest Reconnaissance:

Wheeler had probably the hardest time of any member of the Expedition, and his success in achieving single-handed the mapping of 600 square miles of some of the most mountainous country in the world is sufficient proof of his determination and grit.

The tragic events of 1924 ensured that the early mountaineering history of Mount Everest would be dominated by the mystery of Mallory and Irvine's disappearance on the north ridge in 1924. Meanwhile Oliver Wheeler went on to have a glittering career, rising through the ranks to occupy the same position as Sir George

View from Windy Col Camp at 22,500 feet (6450 metres), showing Everest, North Col and North Peak looking west, 1921.

Everest one hundred years before, becoming Surveyor-General of India from 1941 to 1947. He helped produce thousands of maps in the Allied war effort in the Far East during the Second World War and later retired to his native Canada where he became President of the Alpine Club of Canada from 1950 to 1954.

Members of the 1921 Everest reconnaissance mission. Standing (l-r): Sandy Wollaston, Charles Howard-Bury, Alexander Heron and Harold Raeburn. Sitting (l-r): George Mallory, Oliver Wheeler, Guy Bullock and Henry Morshead.

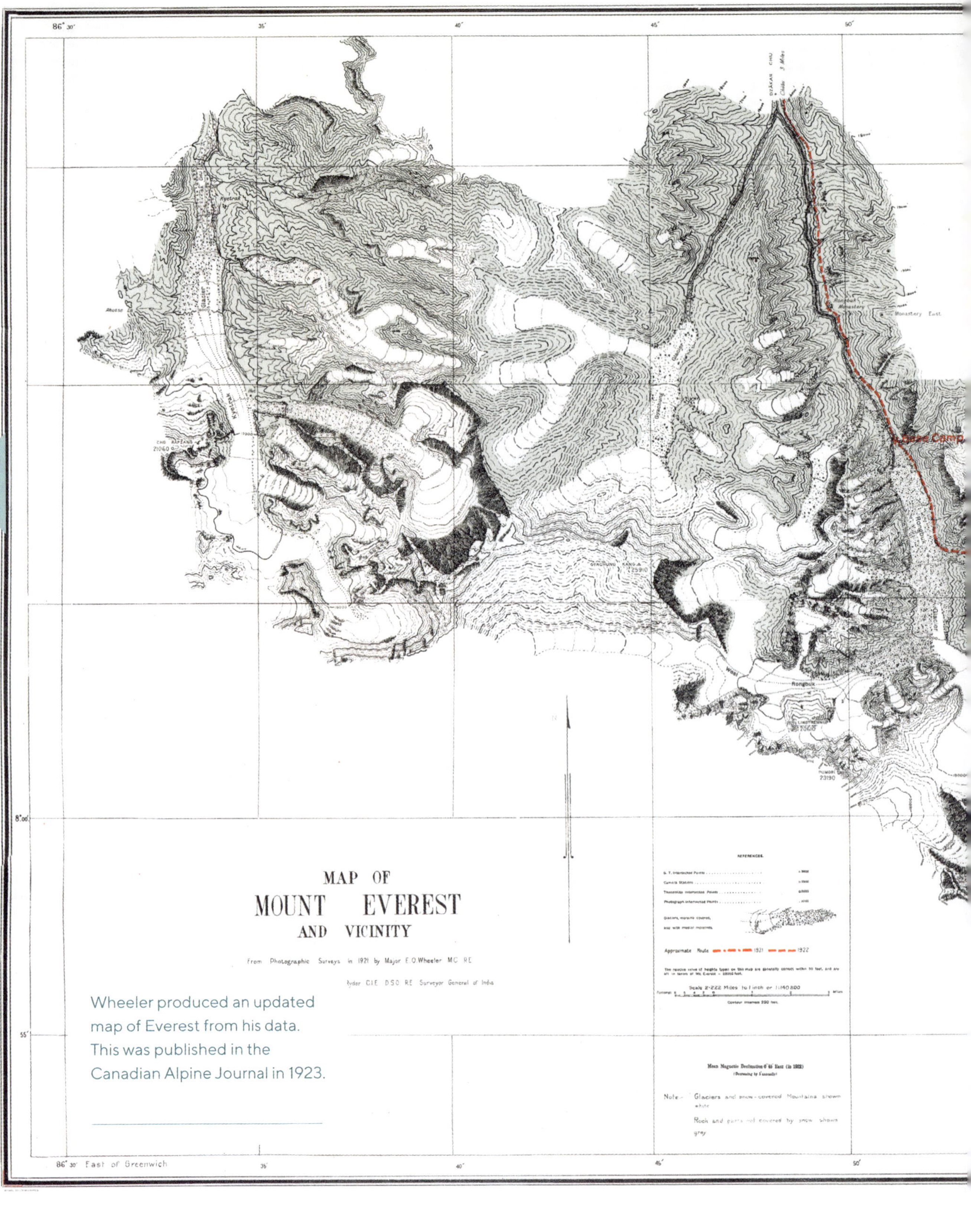

Wheeler produced an updated map of Everest from his data. This was published in the Canadian Alpine Journal in 1923.

Endnotes

1 The words of Sir Francis Younghusband in the introduction to the official report of the 1921 Reconnaissance Mission.
2 Sir George Everest was against the idea of his name being used: he had nothing to do with the mountain's discovery, nor did his name slip easily off the tongue when pronounced in the local language (in fact, in English his name was pronounced Eve-rest, rather than Ever-rest). It is said that the name was used as a compromise to avoid confusion owing to the multiple local names for the mountain.
3 At the time, Younghusband was also president of the RGS. He was a larger-than-life figure with a highly chequered military career behind him and was no stranger to Himalayan adventure himself, having been one of the key players in the Great Game (see elsewhere) during the second half of the 19th century.
4 According to his account in the 1923 Canadian Alpine Journal, these were the instructions that Oliver Wheeler had been given.
5 Wheeler was amongst several members of the Everest team who had served on the Western Front, an episode that is described so poignantly in Wade Davis' *Into the Silence*.
6 Kellas was a scientist who had studied the physiology of humans at high altitude. He believed that, with the right physical constitution and mountain conditions, Everest could be ascended without supplementary oxygen.
7 After producing the first map of Everest in 1921, Wheeler himself would take no further part in the next two Everest expeditions.

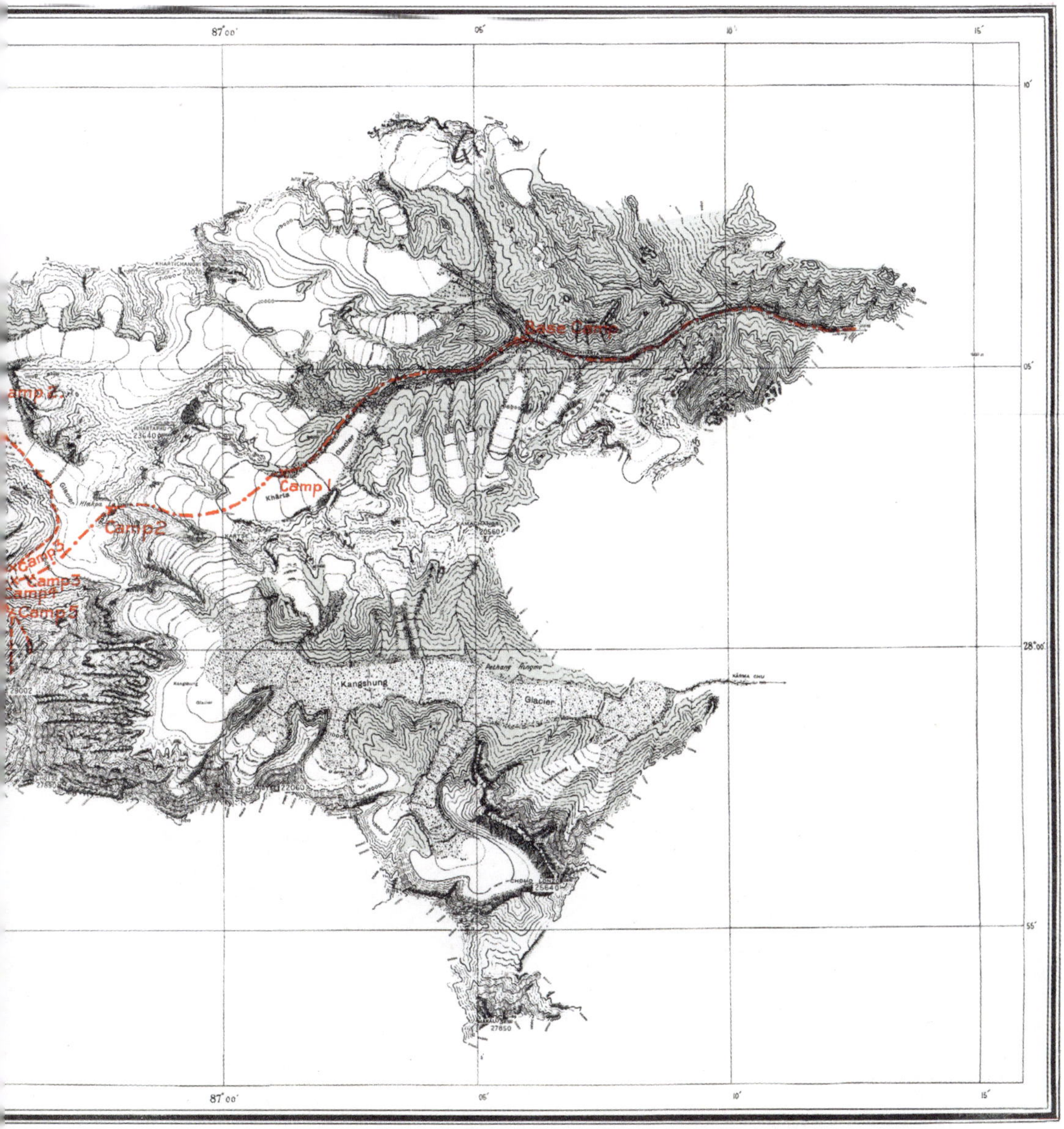

KUH I SANTALLAK ?
ins Muksu-Tal
ins Muksu-Tal
ins Muksu-Tal
PIK GARMO
GARMO-BECKEN
ins Garmo-Tal
FED. WEST

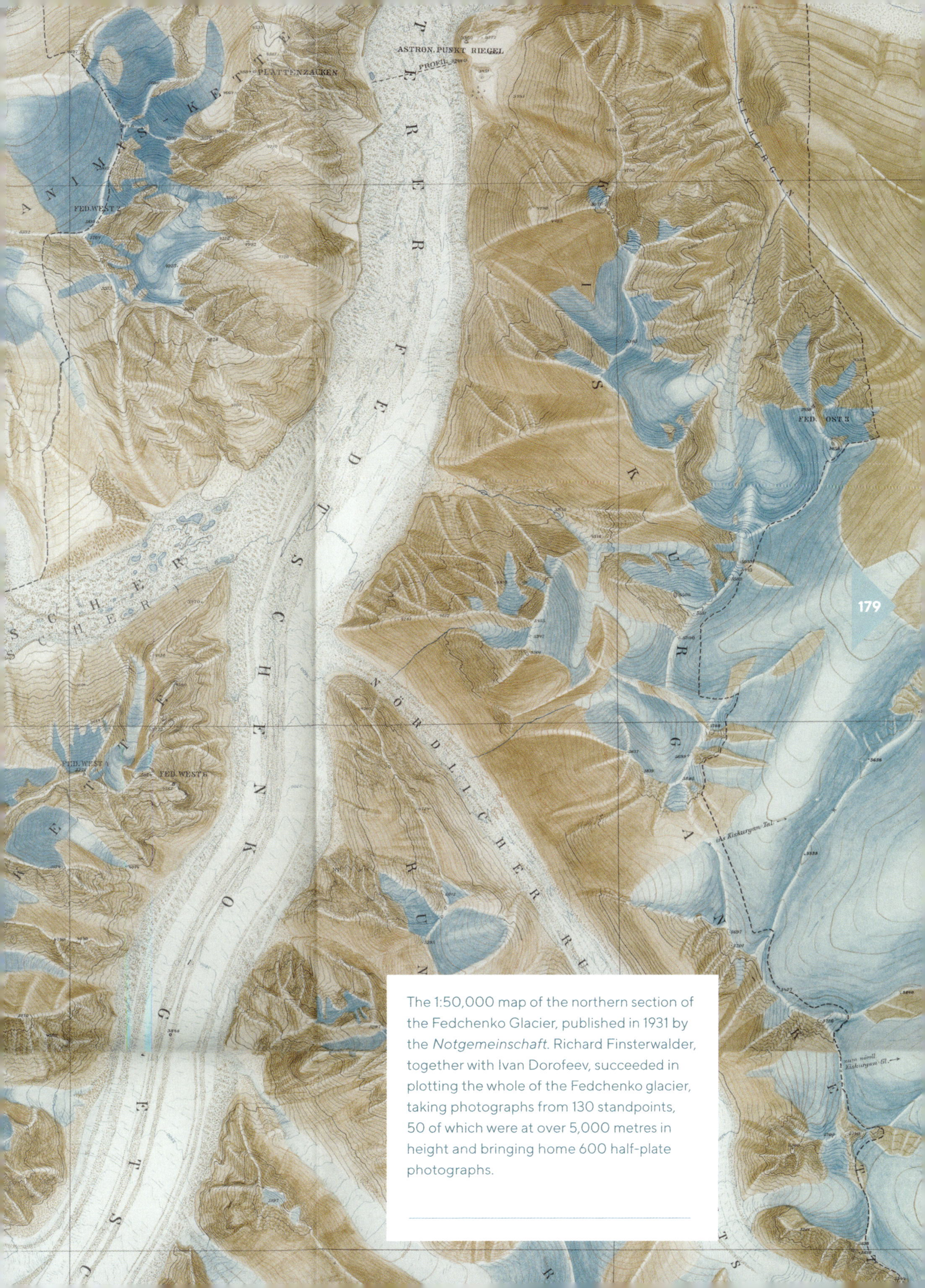

The 1:50,000 map of the northern section of the Fedchenko Glacier, published in 1931 by the *Notgemeinschaft*. Richard Finsterwalder, together with Ivan Dorofeev, succeeded in plotting the whole of the Fedchenko glacier, taking photographs from 130 standpoints, 50 of which were at over 5,000 metres in height and bringing home 600 half-plate photographs.

The slithering giant

How an improbable alliance of Russian and German scientists and mountaineers discovered the true enormity of a subpolar glacier

Snaking its way inexorably northwards through a gargantuan channel for 77 kilometres, the Fedchenko glacier[1] is big on superlatives. This, the longest glacier outside the polar regions, covers an area of over 700 square kilometres, equal to half the size of Greater London. It has a maximum thickness of 1,000 metres, higher than the world's tallest building, and a total volume estimated at no less than 144 billion cubic metres, more voluminous even than the Aswan High Dam on the Nile in Egypt.

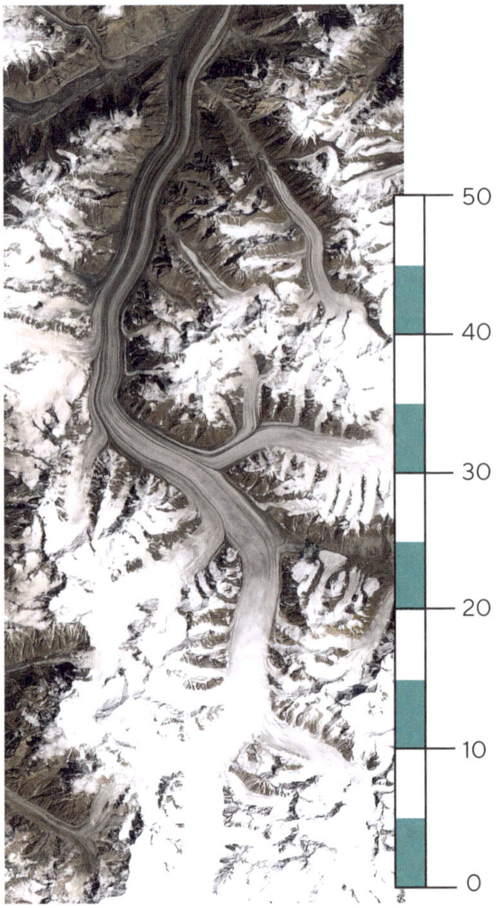

The glacier lies at the heart of the Pamir mountains[2] and is fed by the precipitation which falls on the highest peaks in the western half of the range. Knotted to the Tian Shan in the east and the Hindu Kush in the south, the Pamir forms a node in a complex of immense mountain chains stretching from the Eastern Himalaya in Burma to the Tian Shan on the borders of Kazakhstan and north-west China.

With a combination of east-west and north-south ranges, the mountains of the Trans-Alai (east-west) in the north of the Pamir culminate in Pik Lenin at 7,134 metres. Further south, the Fedchenko slices through the High Pamir, flanked on its western edge by the Akademii Nauk range (north-south) which contains the highest peaks, reaching 7,495 metres at Pik Ismoil Somoni, formerly known as Pik Kommunizma.

Fedchenko Glacier.
Photo Lentz Wolfgang. 1928.

The meltwaters disgorged at the snout of the Fedchenko converge downstream with other river systems spilling down from the High Pamir to form the mighty Amu Darya River (or Oxus). The river reaches no coastline, its waters instead being emptied into an inland basin holding the Aral Sea, thousands of kilometres from the nearest ocean. In fact, the Pamir region is so remote that the rivers it gives rise to all drain into these so-called endorheic basins, with no oceanic outlet. Helped over the millennia by intelligent water management and ingenious water distribution systems, their life-giving waters nevertheless gave rise to several major civilisations in Central Asia.

Because of their complex geography, the thinly populated Pamir mountains have always been difficult for outsiders to penetrate. Their inner sanctum remained unexplored by foreigners until the arrival of the Great Game in the middle of the 19th century, even though their existence was known due to their proximity to the ancient Silk Road. This 'game' was a race for territorial domination by its two adversaries, Russia and Britain, a period that was characterised on the one hand by colonial expansion and on the other hand by exploration, with the two often operating in parallel. Ultimately it was Russia who held the upper hand in the Pamir region and it was assimilated into Turkestan as part of the Russian Empire, and later the Soviet Union.

The Fedchenko glacier itself was only 'discovered' in 1878 by an expedition led by Russian explorer Vasily Oshanin, and even he was oblivious to its true size. He decided to name it after his late friend and fellow explorer Alexei Fedchenko, a scientist who had reconnoitred the Pamir several years before, and in the process discovered Pik Lenin (without otherwise having ever set foot on the eponymous glacier).[3]

The remoteness of the region meant that by the end of the 19th century no surveyor had set sight on the High Pamir, nor had anyone any notion of the magnitude and orientation of the Fedchenko. In the preceding decades, the Russian military had been in such a rush to map the empire that they had simply ignored the desolate mountain regions that stood in their way.

This changed in 1928, when a Russian-German team was sent on a multi-purpose expedition. One of its goals was to map the high peaks and glaciers of the Pamir. This joint undertaking was co-sponsored by the Russian Academy of Science and the *Notgemeinschaft der Deutschen Wissenschaft* ('Emergency Association of German Science'). From a modern perspective these might seem a strange pair of bedfellows only ten years after the countries had been on opposing sides in the First World War. However, in its aftermath, both the Weimar Republic and the Soviet Union found themselves largely ostracised by the international community and gradually gravitated towards each other.

The goals of the mission, which would be officially known as the 'Russo-German Alai-Pamir Expedition 1928', were scientific, and the expedition team comprised eleven Russian and eleven German members. The Russian scientists took charge of mineralogy, petrography and strategically important geodetic-astronomical tasks. The Germans, meanwhile, were responsible for geology, surveying and mapping, glaciology and language research. Overall charge of the expedition was assumed by the Director of the Council of People's Commissars, Nikolai Gorbunov. Organisational leadership was in the hands of the German Willi Rickmer Rickmers, who, together with his compatriots, would be largely responsible for its cartographic and mountaineering aspects.

After studying botany in Vienna, Rickmers developed an avid interest in climbing and exploration. By the age of 45 he, along with his British-born Orientalist wife Mabel, had become an accomplished traveller in, and an expert on Russian Turkestan.[4] He was familiar with existing Russian and British maps of the Pamir of the late 19th century and, as far as high mountain mapping was concerned, he clearly felt that he was on the cusp of a new era, with an imminent transition from discovery to scientific investigation.

Of past cartographers he wrote:

It is difficult to distinguish fact from fancy because the old surveyor was possessed by a horror of the void. His aesthetic sense apparently revolted against white patches. These he peopled with contorted caterpillars created partly from hearsay, partly from his inner consciousness.

By the early 20th century, surveyors had taken on a different role, he believed:

... the world is discovered, that is to say, known everywhere in outline. Bold sketching no longer helps; the mosaic between the contours has to be filled in. The finder is being replaced by the examiner, the prospector by the sinker of shafts, the storyteller by the measurer and the statistician.'

Rickmers had already had one go at surveying the High Pamir as leader of a German and Austrian Alpine Club (DuÖAV) expedition to the so-called Peter the Great Range in 1913. This expedition had omitted the environs of the Fedchenko glacier but was notable because it used the relatively new surveying technique of photogrammetry. It took another 15 years after an intervening World War and a slow return to diplomatic normality before Rickmers would get another bite at the cherry.

Chief surveyor on the 1928 expedition was the 29-year-old Richard Finsterwalder, who had graduated as a surveyor from the Technical University of Munich in 1922.[5] He soon found a job as a cartographer at the DuÖAV, at the cutting edge of mountain mapping at the time, not least through the application of advanced surveying techniques.

The mountaineering team – all members of the DuÖAV – consisted of the Germans, Phillip Borchers, Eugen Allwein, Karl Wien and Franz Kohlhaupt and the Austrian, Erwin Schneider. Rickmers described them as the pick of contemporary alpine climbers, who – in addition to providing scientific assistance – were mainly there for *"geographical patrol work and voluptuary peak-bagging"*.

Willi Rickmer Rickmers (left) leader of the 1928 Pamir expedition, with William Reinig (zoologist) and Ludwig Nöth (geologist).

After a five-day train journey from Moscow, the expedition started out from the city of Osh, on the ancient Silk Road in present-day Kyrgyzstan. From here they crossed the high-mountain desert of Markansu and then rounded the expansive waters of Karakul Lake (said to occupy an old impact crater) at a height of 4,000 metres before penetrating the High Pamir from the east.

For the purposes of exploring the high peaks and glaciers of the High Pamir, Rickmers intended to set up base camp in the upper Tanimas valley, whose eponymous river flows due east before bending south and joining other headwaters flowing into the Amu Darya. Rickmers believed that the valley would unlock the gateway to the 'Seltau'[6] range: previous maps had shown the head of the valley filled by the largest glacier in the region descending from the mythical Tanimas Pass.[7]

Detail of the Notgemeinschaft Glacier, now known as the Grum-Grzhimailo glacier, was also mapped by the 1928 expedition.

Two climbers (Kohlhaupt from Germany and Russian team member Perlin) were sent ahead to locate the glacier and the pass, which the team expected to lead to the valley of the Vanch, a westward flowing river on the other side, that Rickmers had surveyed in 1913 and which also joined the Amu Darya.

When Kohlhaupt and Perlin returned four days later, Rickmers received unexpected news:

Their greatest surprise was the absence of the Tanimas (main) Glacier. Instead of that they found three right side glaciers forming pinnacled barricades across the valley. Looking west, from the top of the pass they saw before them an expanse of snowfields so vast and so level that in the short time at their command they could not quite make out whence they flowed and whither they were bound.

Without realising it, they had stumbled upon the upper reaches of the Fedchenko glacier.

From here, the various members of the expedition went their separate ways for their pre-arranged tasks. The German topographers and climbers explored the whole basin of the upper Fedchenko, making the first ascents of prominent peaks in the vicinity. The Fedchenko and the Notgemeinschaft are divided by a group of mountains at the southern end of the Pamir, now known as the Yasgulem Range, from which both glaciers rise. The range's highest peak, likely nameless since the dawn of time, was christened by the German mountaineers as Dreispitz[8] and was measured accurately at 6,950 metres. The slightly lower Breithorn (6,850 m) was scaled by team members Allwein, Schneider and Wien.

A second fabled pass, separating the Yasgulem and the Academy of Science Ranges, was 'discovered': this was the Kashalayak which had occasionally been used by locals as a high-altitude passage from the Fedchenko to the Vanch Valley.

On 8 September, Rickmer decamped from the Tanimas Valley and rendezvoused with the climbers and topographers in the small settlement of Altyn Mazar, at the foot of the Fedchenko Glacier. The other team members had meanwhile traversed the glacier's length and understood its true size. By then, Finsterwalder had succeeded in plotting the whole of the Fedchenko glacier, with all its branches, and was later able to produce a map at the scale of 1:25,000.

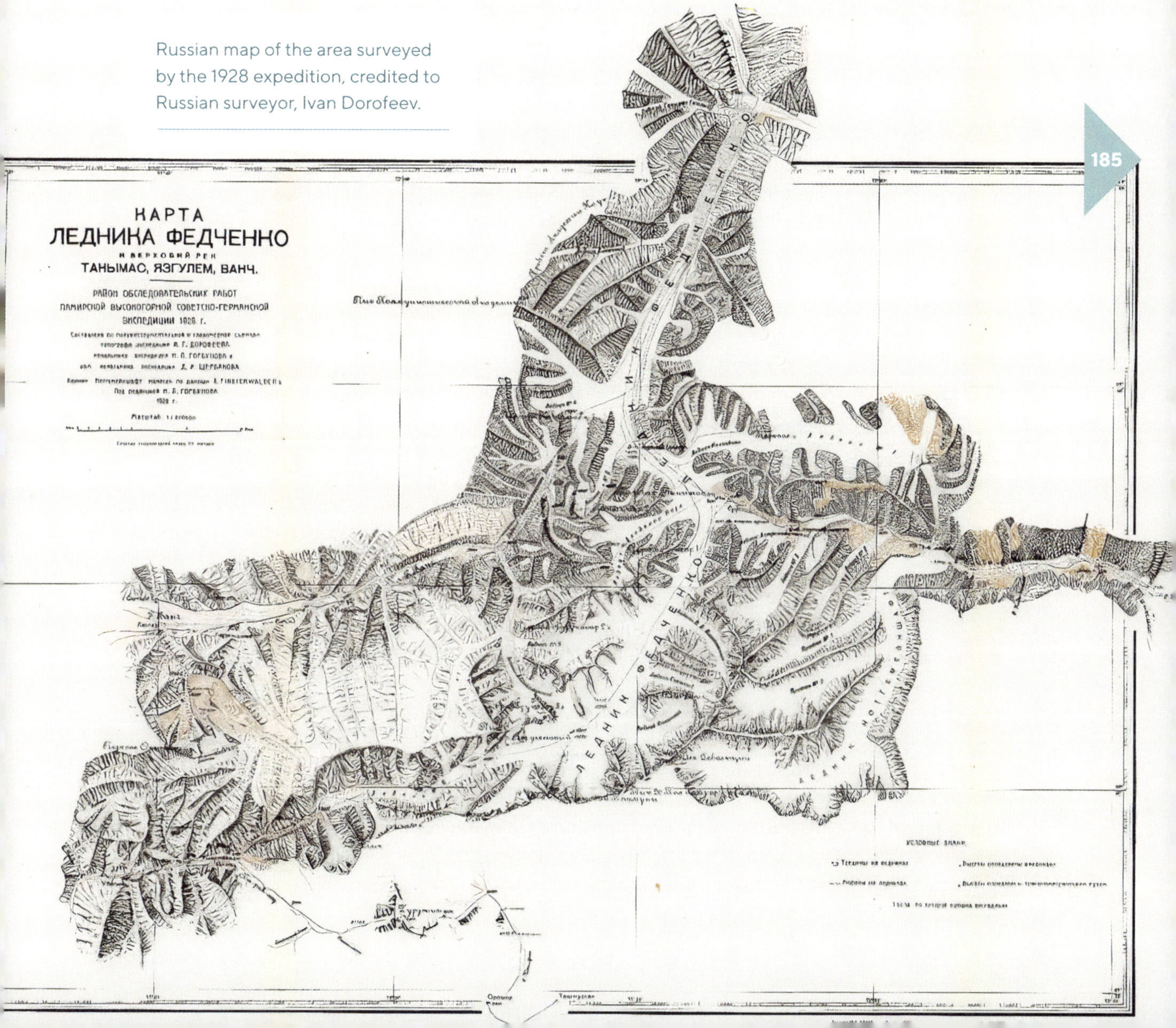

Russian map of the area surveyed by the 1928 expedition, credited to Russian surveyor, Ivan Dorofeev.

With his Soviet counterpart Ivan Dorofeev, he photographed from 130 standpoints, 50 of which were at over 5,000 metres in height, bringing home 600 half-plate photographs. He also measured various baselines and made a series of triangulations, which were calibrated with astronomical readings taken by professor Ivan Belyaev of the Russian team.

It was from this base at Altyn Mazar that three climbers of the DuÖAV set out eastwards on 18 September to make an attempt on Pik Kaufmann (7,134 m) – later redubbed Pik Lenin – in the Trans-Alai Range. Russian explorer Alexei Fedchenko had identified this peak 57 years previously as being the highest in the range and at the time of the 1928 expedition. It was believed to be the highest summit not only in the Pamir, but also in the Soviet Union.

Eugen Allwein, Karl Wien and Erwin Schneider trekked upstream along the Saukdara valley on the southern flanks of the Trans-Alai Range, until they could no longer tread on solid ground. From the head of the valley, they had to ascend the Greater Saukdara Glacier towards a saddle known as the Krilenko Pass at an elevation of 5,820 m. After climbing from the saddle along the North-east Ridge they reached the summit in the afternoon of 25 September. Having reached the top of Pik Kaufmann, they had succeeded in scaling what was believed to be the highest point in the Pamir, but they had also broken the altitude record for the highest ascent of any summit in the world. For the summiteers, this would have surely represented not only the icing on the cake as far as the expedition was concerned, but also the height of their mountaineering careers thus far.

Overall, the joint Russo-German mission was deemed an outstanding success and its findings – not just with respect to mapping and mountaineering, but to geology, glaciology, mineralogy and meteorology too – were published in five volumes of text and one volume of maps. The general view among participants was that collaboration between mountaineers and researchers was productive, and morale amongst team members had been good. Rickmers' own verdict:

It is almost a law that large travelling parties never come off without a fight. But in this case a miracle was wrought, namely that around thirty men, both Russians and Germans, lived side by side in harmony for several months, even if a cathartic word may occasionally have been spoken.

The 1928 Alai-Pamir Expedition was successful in almost every respect, with perhaps just one exception: it failed to conclusively establish the identity of the highest peak in the Pamir. In the south, the Dreispitz, at 6,950 metres, had been singled out by the expedition as the highest. Later, when photogrammetric records of the 1928 expedition had been analysed, it was found that one of the peaks visible to the west of the Fedchenko glacier reached a height of 7,495 m. It was left to a Soviet expedition of 1933 to determine definitively that this elusive mountain (which they referred to as 'Pik Garmo') was indeed the highest in the Pamir. It was during this later expedition, mounted by Nikolai Gorbunov, that the first ascent of the mountain was made by Russian mountaineer Yevgeniy Abalakov. In true Soviet style, being the loftiest mountain in its dominions, Pik Garmo[9] was rechristened Pik Stalin and then – during the de-Stalinisation process of the 1960s – it became Pik Kommunizma (or 'Communism Peak'). Needless to say, its name was once again changed after the collapse of the Soviet Union and its incorporation into Tajikistan: Pik Ismoil Somoni.[10]

The 1933 expedition was not, Gorbunov contended, just a climb or a matter of sport, but a scientific undertaking which aimed to place portable weather stations as high up as possible and signalled the start of further glaciological, hydrological and meteorological research by the Soviets in the High Pamir. In 1935, they built a permanent hydrometeorological station halfway up the Fedchenko Glacier at 4,169 m, known as the Gorbunov Meteorological station.[11] Remarkably, the station continued operating for almost 60 years, but was abandoned in 1992. Maintenance and provisioning of this high-altitude research station was always a huge logistical undertaking and it was the break-up of the Soviet Union and the ensuing civil war in Tajikistan which ultimately marked its demise.

As for the Fedchenko, which continues on its slow, silent and seemingly interminable journey, barring the scant few mountaineering expeditions and trekkers that pass this way, it remains as remote and desolate today as it ever was.

Endnotes

1 In 2023, Tajikistan officially renamed the glacier as part of a derussification programme. It is now known officially as as the Vanch-Yakh Glacier.
2 The name Pamir is thought to have been derived from the word for the long, flat valleys dividing these principal ranges and which were often used for summer pasturing by the local Tajik people.
3 Fedchenko died in 1873 at the age of 29 on Mont Blanc, where he had spent the summer studying glaciers in preparation for another expedition to Central Asia.
4 Roughly equivalent to the area covered by Tajikistan, Kyrgyzstan, Uzbekistan, Turkmenistan and southern Kazakhstan today.
5 His father Sebastian Finsterwalder was a mathematician and glaciologist and taught at the same university. He is generally acknowledged as having introduced glacier photogrammetry as a way of determining the shape of glaciers. Richard Finsterwalder produced the first detailed map of Nanga Parbat at a scale of 1:50,000 following the fateful German expedition to climb the peak in 1934. He went on to have an illustrious career as a geodesist, glaciologist and mountain cartographer.
6 Seltau was the name that Rickmers used to refer to the High Pamir.
7 The Tanimas was not the only legendary pass that was said to cross the High Pamir. Another was the Kashalayak. They were said to have been used in ancient times by natives and had either become obsolete due to an increase in ice impeding their access or they had been kept secret by the local Tajiki to hide their existence from their northern rivals the Kyrgyz for the purposes of flight or for the supply of reinforcements for battle.
8 Christened *Pik* Revolyutsii by the Soviets in 1945 and rechristened Qullai Istiqlol (or 'Independence Peak') in 2006
9 To confuse toponymical matters further, the name Pik Garmo was eventually assigned to a lower peak (6,595m) in the south of the Academy of Sciences range, at the head of the Garmo Glacier.
10 Ismail Samani was the ruler of the Samanid dynasty which became a powerful force in Central Asia towards the end of the first millennium.
11 Nikolai Petrovich Gorbunov (1892 – 1938) was a Soviet politician, chemist, engineer and academic and one-time personal secretary to Lenin.

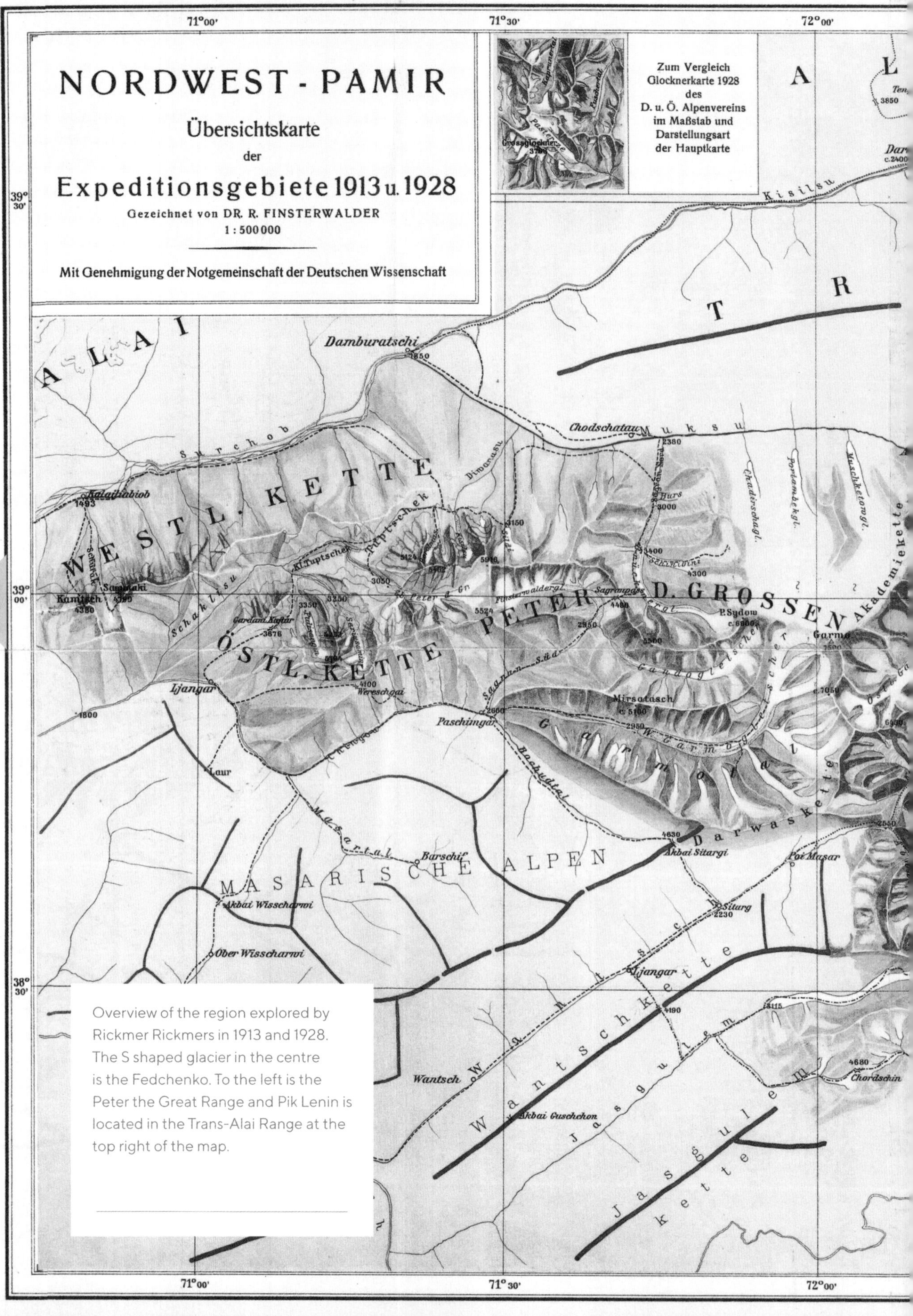

Overview of the region explored by Rickmer Rickmers in 1913 and 1928. The S shaped glacier in the centre is the Fedchenko. To the left is the Peter the Great Range and Pik Lenin is located in the Trans-Alai Range at the top right of the map.

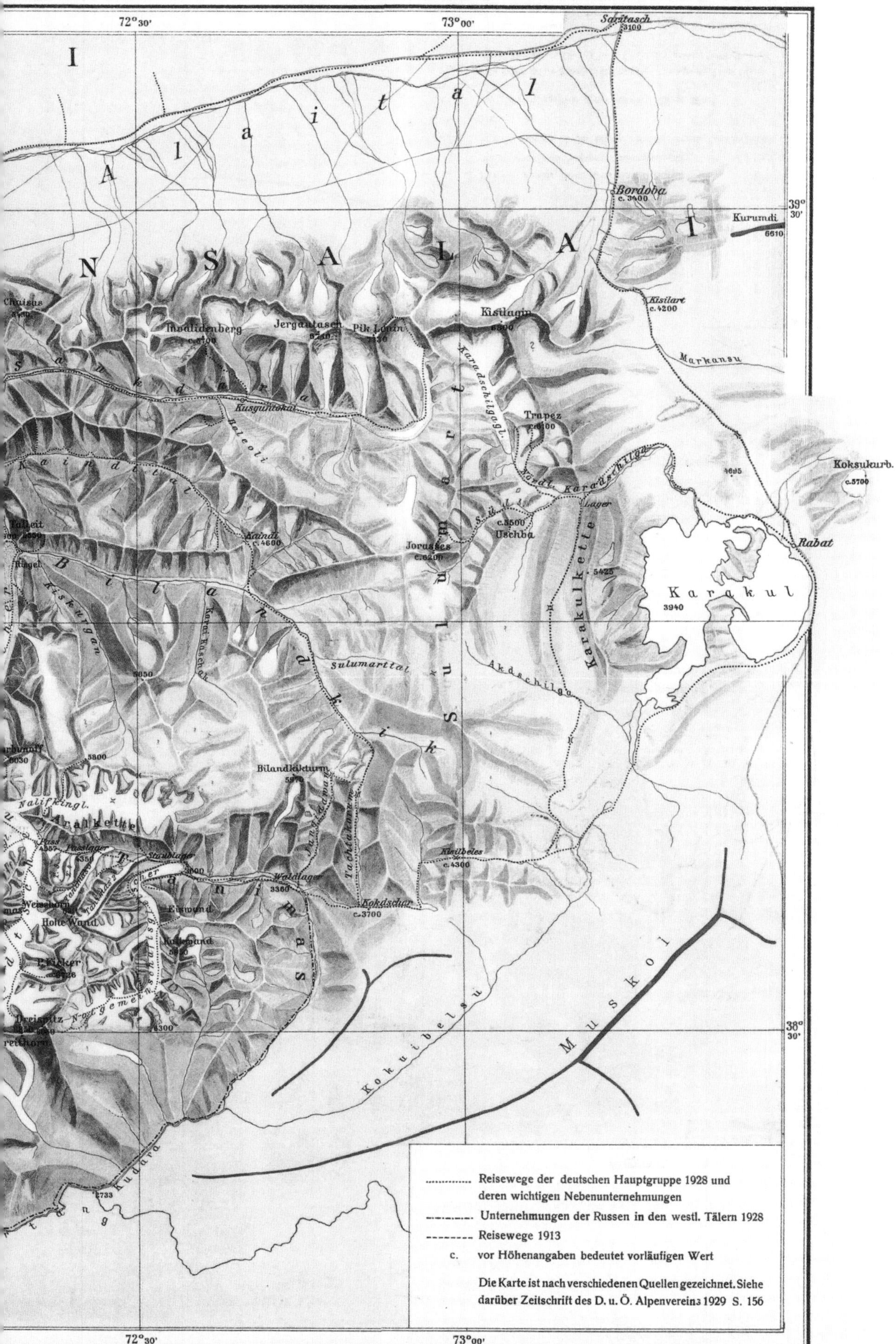
72°30'
73°00'
Saritasch
3100
Alaital
Bordoba
c. 3400
39°
30'
Kurumdi
6610
TRANSALAI
Kisilart
c. 4200
Markansu
Kisilagin
6800
Jergautasch
Pik Lenin
Kusguntokai
Karadschilgagl.
Trupez
c. 6100
Koksukurb.
c. 5700
4695
Nördl. Karadschilga
Lager
c. 5500
Uschba
Jorasses
c. 6200
Rabat
Kainditai
Kaindi
c. 4600
Karakul
3940
5425
Karakulkette
Sulumarttal
Akdschilga
5850
5800
Bilandkiikturm
Nalifkingl.
Kisilbeles
c. 4300
Staublager
Waldlager
3360
Kokdschar
c. 3700
Weisshorn
Holte Wand
Eiswand
Kokuibelsu
Muskol
38°
30'
Kudara
2733
Reisewege der deutschen Hauptgruppe 1928 und deren wichtigen Nebenunternehmungen
Unternehmungen der Russen in den westl. Tälern 1928
Reisewege 1913
c. vor Höhenangaben bedeutet vorläufigen Wert
Die Karte ist nach verschiedenen Quellen gezeichnet. Siehe darüber Zeitschrift des D. u. Ö. Alpenvereins 1929 S. 156

Swiss-style
Snapshots from a nation's cartographic past

Like watchmaking and railway engineering, but perhaps not quite so self-evidently, there was a general acknowledgement in the wider world that cartography was something that the Swiss were good at. When mountaineering became a popular sport in the 19th century, the newly formed alpine clubs around Europe instigated their own mapmaking services to redress the inadequacies of the mountain maps produced by their own national mapping agencies.

Towards the end of the 19th century, the Swiss Alpine Club (SAC) had the sense to realise, particularly in the wake of the Dufour and Siegfried maps, that the Bern-based *Eidgenössische Topographische Bureau* was already turning out cutting-edge cartography, and for a while the SAC even subsidised production of the Siegfried map. Often, field surveyors employed by the *Bureau* and who, in the course of their duties acquired an intimate knowledge of mountain terrain (for example, Xaver Imfeld), were themselves active members of the SAC. After the First World War, the agency started introducing photographic processes to their work using stereoscopic photography. It lessened the burden of field surveying and meant cartographers could do the bulk of their work back in their offices.

Thanks largely to their expertise in mountain topography, Swiss cartographers were also highly sought after by foreign institutions looking for a good mountain map. One such cartographer was Charles Jacot-Guillarmod (1868-1925), whose early career followed a similar path to that of Xaver Imfeld: he worked for the *Eidgenössische Topographische Bureau* for 24 years between 1890 and 1914, where he too acquired a speciality in the drawing of rock features. Shortly after his dismissal[1] from the agency in 1914, he went on to produce the first-ever detailed maps of K2 and Kanchenjunga, without ever stepping foot in the Himalayas. Of the two, the K2 map was particularly accurate and is based on information from the Duke of the Abruzzi's expedition to the Karakoram in 1909. Amazingly, Jacot-Guillarmod, from the confines of his office, was able to crunch data which had been recorded at a number of carefully calculated triangulation stations set up by the Duke's surveyors. Alongside high-quality photographic material from Vittorio Sella, the expedition's photographer, he was able to add rock drawings and contour lines to the ensuing map.

The accuracy of the Kangchenjunga map is more questionable because he was working with imprecise location data and photographs (likewise taken by Sella, in 1899). Uncannily, both maps demonstrate the 'Swiss Style' that was (and still is) manifest in the *Bureau*'s maps.[2] His most significant work was his Everest map, which was commissioned through Jacot-Guillarmod's publisher in Switzerland by the Royal Geographical Society (RGS) in 1925.[3] This was to be an updated map of Oliver Wheeler's Everest map based on the photographic material and surveying data he collected in 1921 and subsequent data from the two subsequent British expeditions. Jacot-Guillarmod spent several

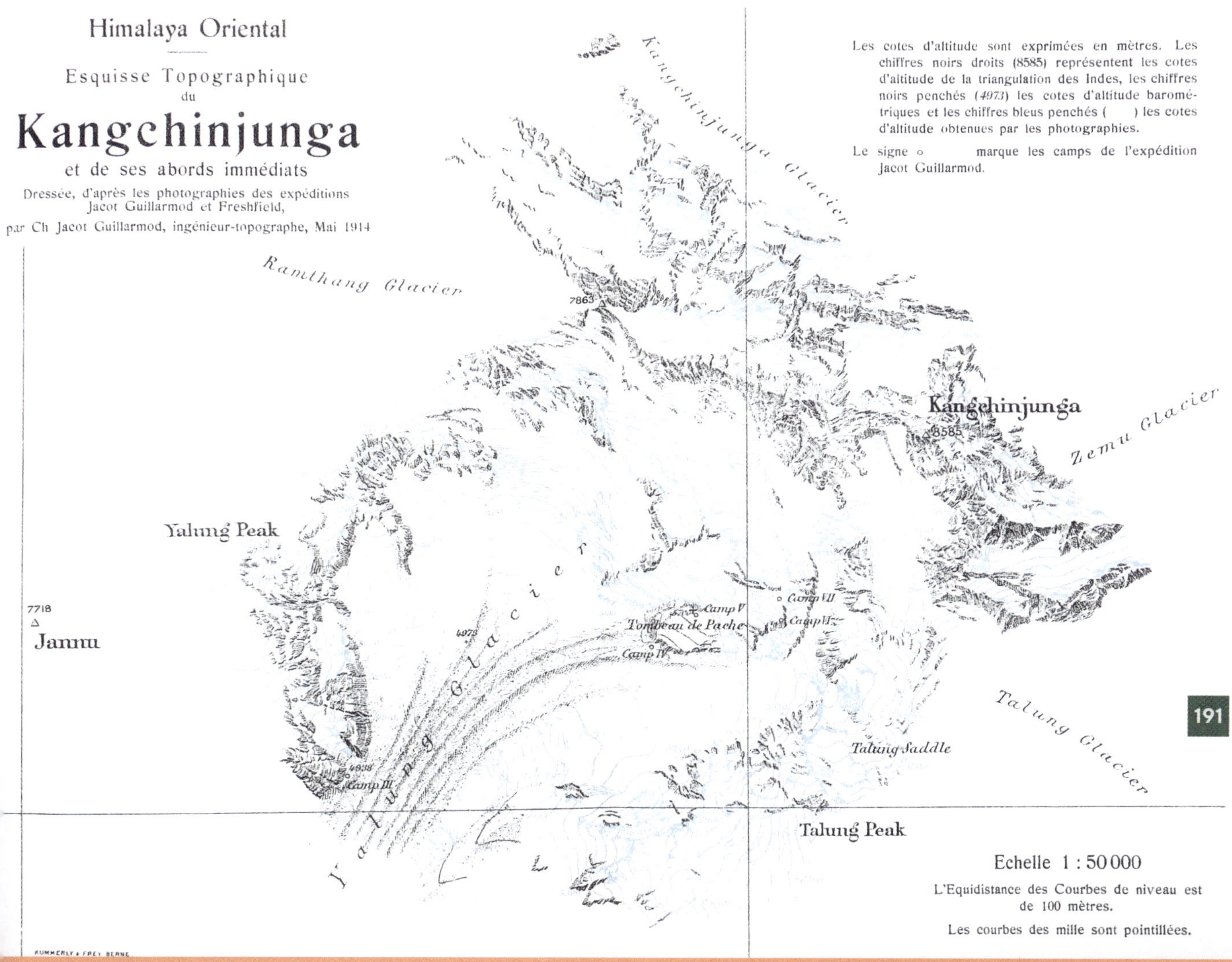

Without ever setting foot in the Himalayas, Charles Jacot-Guillarmod created this 1914 'Swiss-style' map of Kanchenjunga using data that was derived from sketchy location data and photographs taken by Vittorio Sella on an 1899 expedition to the mountain.

weeks in London working on the map before returning to Switzerland to complete the work which was published later in the year at a scale of 1:63,360. Sadly, Charles Jacot-Guillarmod never got to see the finished Swiss-style version of the Everest map: he passed away on 14 August 1925 just weeks before its publication.

In the 20th century, few people can have had such an impact on relief depiction worldwide as Eduard Imhof (1895–1986). Born in Schiers in the canton of Graubünden, Imhof was immersed in Swiss topography from an early age: his father was a Geography teacher who had a sideline writing travel guides to the local mountains for the SAC, and it was on expeditions with his father that the young Eduard must have first become acquainted with the Siegfried maps. Both artistically and technically gifted, he chose to study geodesy at the ETH in Zürich. In his undergraduate days he frequently went for multi-day excursions in the Bernese Oberland and Valais. At the ETH, he was taught by Professor Fridolin Becker[4] (1854–1922), credited with introducing shaded relief maps, who stirred in Imhof a passion for relief presentation.

When Becker became ill in 1919 and could no longer perform his teaching duties, Imhof deputised for him despite having only recently graduated and then went on to replace him when his mentor died. From 1925 to 1965, Imhof was professor of cartography at the institute, which included surveying and map drawing. Relief presentation – essentially making a two-dimensional surface look three-dimensional – was a field in which he more than excelled. He perfected oblique illumination and introduced hypsometric tinting to his maps, a technique which attempted to mirror the topography on the map to its appearance from an aircraft at a high altitude. Looking down from this great height, the lower regions appear hazier in a grey-blue tone, the hills and middle ranges vary from blue-green to green and yellow-green, while the peaks, more exposed to sunlight in the thinner air, are in shades closer to white. His approach, which had evolved and been refined since the time of Becker, was of course based on the Swiss Style. Not only did Imhof manage to persuade the Swiss federal mapping agency to introduce his techniques for the new topographic maps at 1:25,000 and 1:50,000, his methods were also adopted by other agencies worldwide. From 1932 to 1976, he was also put in charge of editing the Swiss *Mittelschulatlas*, used in schools throughout the country. First published in 1965, his book *Kartographische Geländedarstellung*, later translated in English under the title of *Cartographic Relief Presentation* in 1982, which explained his philosophy of mapping, became the go-to *vade mecum* for topographic cartographers.

In the second half of the 20th century, the *Bundesamt für Landestopografie*, today known as swisstopo, enjoyed collaborations with the US National Geographic Society and Bradford Washburn. The latter was an American mountaineer and photographer, who later turned his hand to cartography in his capacity as director at the Boston Museum of Science (see chapter 18). In the late 1950s, Washburn undertook a project to map Mount McKinley (now Denali) in Alaska. From photographic reconnaissance of the massif, he realised that to meet his demands for detail, accuracy and aesthetics, particularly for rock presentation, special skills were needed, so he insisted on using Swiss cartographers. The map was eventually published in 1960 and has all the hallmarks of a swisstopo map. So pleased was he with the results of these labours, he went on to perform two other major projects with cartographers from the Swiss mapping agency: the Heart of the Grand Canyon map, published in 1978 by the National Geographic Society; and probably his magnum opus, the 1:50,000 map of Mount Everest in 1988 – another joint collaboration – which marked the National Geographic Society's centenary and the 150th anniversary of the Swiss Federal Office of Topography.[5]

Of his Mount McKinley map, Washburn was asked, *"Why on earth are you having your map published in Switzerland?" "It is my conviction,"* he replied, *"that no other country in the world could equal the job done by the* ***Landestopographie.****"* Coming from such a distinguished mountaineer and cartographer as Bradford Washburn, it is hard to think of richer praise for Swiss cartography.

* * *

Swisstopo

The Swiss national agency for mapping has undergone several incarnations since it was founded in Carouge, near Geneva, in 1838 as the *Bureau topographique federal / Eidgenössisches Topographisches Bureau.* Since 1979, the organisation responsible for Swiss cartography has been known in German as the *Bundesamt* für *Landestopografie* and in French as the *Office fédéral de topographie*, with its headquarters based in the canton of Bern. Operating in a global, multilingual environment, since 2002, the agency has branded itself simply as *swisstopo*, alongside its official names. For the record, the Swiss Federal Office of Topography (the official English name) is known in Italian as the *Ufficio federale di topografia* and in Romansh as the *Uffizi federal da topografia.*

* * *

Endnotes

1 It is thought he was a victim of workplace bullying.
2 Swiss-style maps, for example, are characterised by their distinctive relief shading. The topography has a vibrant appearance through using a combination of contour lines, rock depictions, 'illuminated' relief shading and colour variation.
3 In correspondence between the RGS and Jacot-Guillarmod, it was reported that cartographers at the RGS were overloaded and so unable to carry out the task.
4 Becker has also worked with Xaver Imfeld on relief models for the *Gletschergarten* in Lucerne.
5 Jacot-Guillarmod's 1925 Swiss-style map is not wholly unrecognisable from the map of Mount Everest published in 1988.

Eduard Imhof was a master of relief presentation. He devised cartographic techniques which aimed to mirror the topography on the map to its appearance from an aircraft at a high altitude, such as this map of the Jungfrau and the Aletsch Glacier in the 1969 *Mittelschulatlas*.

Michael Spender's 1:250,000 map of the area covered by the Shaksgam Expedition.
After spending the summer of 1937 scouring a huge tract of unexplored territory in the Karakoram, Eric Shipton and his fellow expedition members arrived back in civilisation, tired and dishevelled. During the previous three months they had surveyed no less than 4,661 square kilometres of terrain which was home to some of the most awe-inspiring peaks and glaciers on the face of the earth.

HE GREAT KARAKORAM
HE AGHIL MOUNTAINS
ichael Spender and colleagues
aksgam Expedition 1937
Yarkand River
Bazar Dara
Yarkand River
Shaksgam R.
Jangal
Shaksgam or Muztagh R.
Aghil Pass
Jangal
Zug Shaksgam River
Durbin Jangal
Expedition Base Dump
Shaksgam River
Muztagh Pass
Muztagh Tower
7280
Scale 1: 250,000
Miles
Kilometres
The new names on this map are those proposed to the Survey of India by Mr. Spender.
Contours at intervals of 100 metres.
Heights in metres
76°15′
76°30′
76°45′
36°15′
36°0′

Off the map

Under the guidance of Eric Shipton, a lightweight expedition undertook an epic adventure to explore the back of beyond in the Karakoram

After four months of tramping in the wilderness, they arrived back in civilisation besmirched, ragged and unwashed. Their unkempt hair and matted beards aroused even the consternation of the local populace, themselves not unaccustomed to a sparse existence on the edges of the habitable world. As dishevelled as they may have seemed, the two figures who arrived in Shimshal – a back-of-beyond cluster of settlements in Baltistan, northern Kashmir – had actually spent the summer of 1937 scouring a huge tract of uncharted territory, assiduously filling in the blanks on an empty map. It was no ordinary terrain, containing some of the most awe-inspiring peaks and glaciers on the face of the earth.

On his way back from an Everest expedition in 1936, Eric Shipton (1907-1977) devised a plan to reconnoitre the Karakoram in the western Himalaya. Already an accomplished mountaineer at the age of 29, Shipton had several successful Himalayan expeditions under his belt, as well as a number of intrepid treks in the Mount Kenya, Kilimanjaro and Rwenzori ranges of East Africa. Exploration was an integral part of his make-up and he was drawn to the Karakoram simply because it was a void that hadn't yet been mapped.

The Karakoram Range is a tangled knot of mountains that forms part of the watershed between Central Asia and the Indian subcontinent. Until the 1930s, there had been little motivation to map the Karakoram: it is a desolate region, devoid of human habitation, bristling with mountains of immense dimensions which rise up from deep valleys gouged out by some of the most extensive glaciers outside the polar regions. It could quite justifiably lay claim to the title 'roof of the world', since four of its summits, K2, Gasherbrum I, Gasherbrum II and Broad Peak, surpass a height of 8,000 metres. K2, just 238 metres shy of Everest's elevation and

Eric Shipton in Rishi Ganga, India, 1935.

the second highest peak in the world, never had a local name because, quite simply, there were no locals to give it one.[1]

Yet its remoteness did not mean this mountain fortress had been completely untrodden or unmapped. Ancient caravans had used crossing points between Turkestan and Kashmir to trade goods between north and south, before shifting glaciers had eventually made it impassable to traffic in the mid-19th century.

Largely spurred on by geopolitical necessity, Europeans started taking an interest in the region. In 1861, Lieutenant-Colonel Godwin-Austen explored the mountains as part of the Great Trigonometrical Survey of India, and surveyed the glaciers at the base of K2, one of which was later named after him. The first European known to have traversed this mountain range from north to south was Francis Younghusband in 1887. In a venture in keeping with the age of exploration and empire, Younghusband – a British army officer – undertook a journey of epic proportions from Peking, across the Gobi desert and eventually traversing the Karakoram via the Muztagh Pass (5,422 m), the descent of which – without proper mountaineering equipment – had been a heart-stopping, do-or-die enterprise.[2]

The first ever attempt on K2 was made in 1902 by a party led by English mountaineer, Oscar Eckenstein. Jules Jacot-Guillarmod, a Swiss mountaineer and physician who participated as expedition doctor, attained a height of 6,700 metres on the

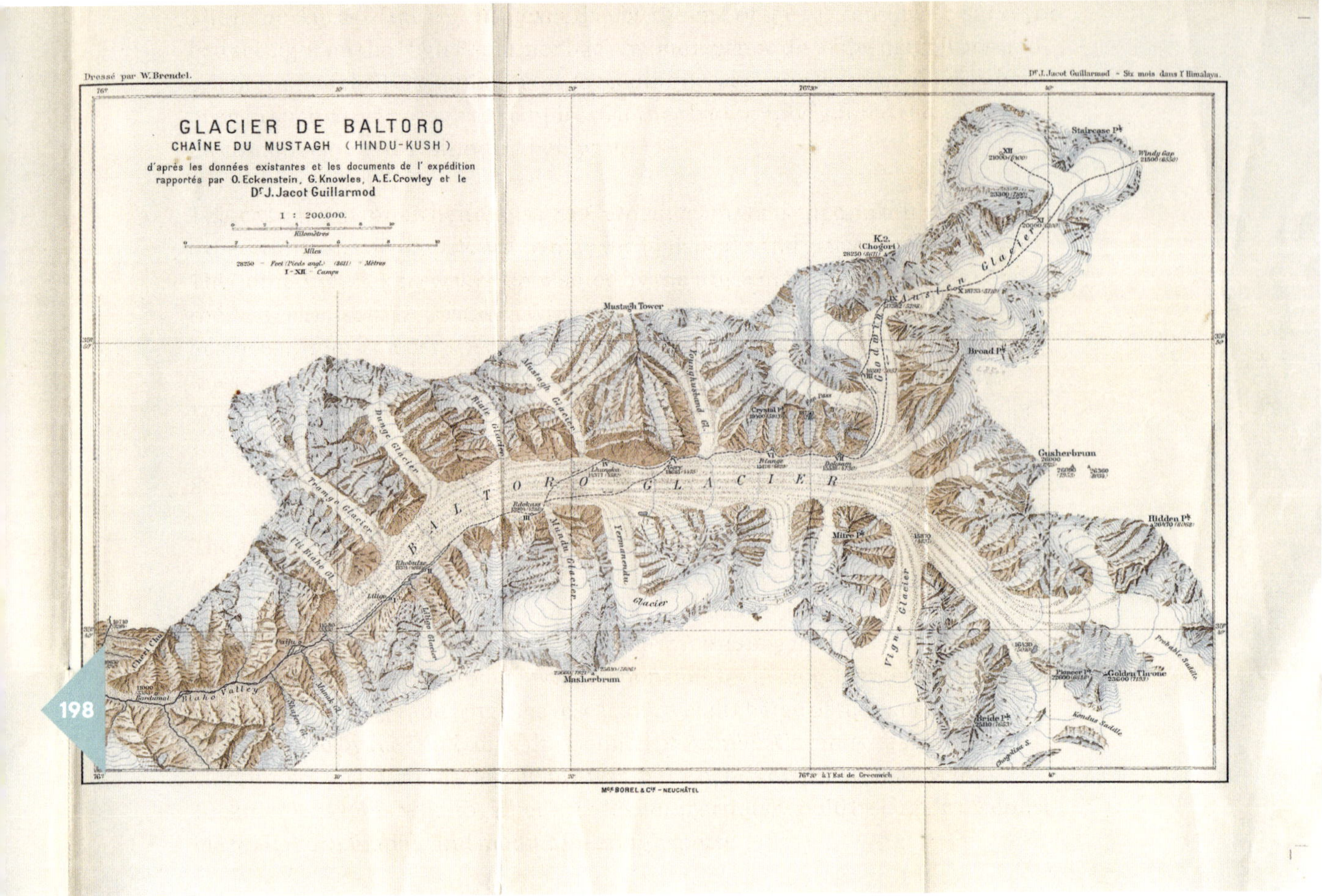

The 1:200,000 map of the Baltoro Glacier surveyed by Jules Jacot-Guillarmod in 1902.

north-east ridge of the mountain. Not only did Jacot-Guillarmod bring home nearly a thousand stereoscopic photographs from the expedition, including the first recorded photograph of K2, he also produced a detailed map of the Baltoro Glacier including the path the party took up the Godwin-Austen Glacier to reach their high point.

Other notable visitors to the Karakoram included the formidable Fanny Bullock Workman. Born into a wealthy American family in 1859, Fanny Bullock married William Workman, a medical doctor ten years her senior, when she was just 22. They spent the best part of the next 30 years travelling the world in search of adventure and were eventually introduced to high-altitude climbing in the Himalayas. In 1906 they made the first ascent of Pinnacle Peak (6,930 m), a women's altitude record at the time. Five years later, along with a team of surveyors, they embarked on an expedition to map the 76-kilometre-long Siachen Glacier in the Eastern Karakoram.

The 1:100,000 map of the upper part of the Baltoro Glacier and K2 produced as part of the Duke of the Abruzzi's 1909 expedition.

People with no first-hand experience of the Karakoram can only imagine the sheer grandeur of the landscape. Few can surely have come as close to capturing its magnificence than Vittorio Sella, possibly the finest and most accomplished mountain photographer of all time, who accompanied an Italian expedition making a (subsequently unsuccessful) attempt on K2 in 1909. The expedition was led by Prince Luigi Amadeo, better known as the Duke of the Abruzzi (see chapter 14), whose explorations, amongst others, included first ascents in Alaska and British East Africa.

By 1937, whilst some parts of the Karakoram – including the southern flanks of K2 – had evidently been surveyed, Eric Shipton set himself the task of filling in some of the gaping gaps north of the main crest. Exploration, more so than peak-bagging, had become his *raison d'être* and his plan came at an opportune time: exploration, for exploration's sake was still *de rigueur*, and the Government of India, under British colonial control, was keen to establish the true extent of its territory for reasons of national security. To achieve this goal, Shipton would need a support team of climbers, surveyors and porters, and – crucially – funding. For the latter, he received money from the Royal Geographical Society, the Survey of India and the Royal Society. He budgeted for a total cost of £855[3] (around £73,000 in current terms).

The lead team consisted of Shipton himself; Bill Tilman (1898-1977), another climber with a complete back catalogue of mountaineering feats, frequently in the company of Shipton and in every way his alter ego; Michael Spender[4], a surveyor who had given invaluable service on an Everest reconnaissance mission led by Shipton in 1935. To all intents and purposes, Spender would be responsible for coordinating the surveying work of the expedition; and John Auden[5], a gifted geologist, who began his career with the Geological Survey of India in 1926. The team was completed by seven Sherpas[6] who had given sterling service as mountain guides on previous Everest campaigns.

The logistics of a four-month trek in inhospitable and uncharted mountain terrain were daunting, not least with heavy and valuable surveying equipment. Shipton hated the massive scale of sponsored expeditions. He was very much a free-spirit, lacking the single-mindedness of an expedition climber focused solely on reaching the top. However, for such an odyssey, the Karakoram project required the services of a hundred local porters to carry food, tents and other provisions.

Roughly speaking, Shipton's plan was:

... to reach the Baltoro glacier by the end of May; to cross the watershed with sufficient food to last the party for one hundred days after reaching its base below the snout of the Sarpo Laggo glacier; to cross the Shaksgam [river] and spend as much time in the Aghil range as possible without being cut off by the summer floods; to return to the Sarpo Laggo about the middle of July and to spend the remaining two months working on our other two objectives, the Crevasse glacier region and the area to the north of K2.

Vittorio Sella's photograph of the Trango Towers in 1909.

Once they had left Askole, Baltistan on 26 May, their first major task - the most critical of the whole expedition - was to find a passable crossing over the watershed to the north of the Baltoro glacier, a broad and sweeping glacier hemmed in on all sides by the Himalayan giants. An Italian survey team working in the Baltoro valley in 1929 had surmised that there was a more permissible, more westerly route across the range to the Sarpo Laggo Glacier on the other side than the vertiginous Muztagh Pass taken by Younghusband in 1887. After reconnoitring the Trago Glacier that occupied a valley branching off to the north of the Baltoro, Shipton and his party eventually found the saddle that gave them access to the Sarpo Laggo Glacier on 2 June. However, this was not without much trial and error, which included mutiny from the porters and fever on the part of Tilman and Auden. Over the course of several days, they ferried their supplies to their expedition base at the foot of the glacier.

Michael Spender plane tabling in difficult conditions, Jammu & Kashmir, 1937

It was here that Spender laid a surveying base in order to get K2 and a number of local points plotted onto his plane-table sheet. K2 would be the fixed point from which measurements would be made during the trip. After several days exploring the terrain in the vicinity of the base camp, the party set off in a northerly direction, retracing the route taken by Younghusband to the Aghil Pass, named after the mountain range to the north of the central massif of the Karakoram. They eventually arrived at the pass on 20 June, its summit identifiable from the description made by Younghusband 50 years previously. From here, they made several sorties into smaller side valleys and also followed the main valley down to its confluence with the Yarkand, a 1,000-km-long river that flows into the Tarim river which eventually dries out in the Taklamakan Desert in the Xinjiang province of China.

Hardly having time to rest at their base at Suget Jangal back in the Sarpo Laggo Valley on their return from the Aghil, Shipton, Tilman and Auden spent a week in the web of valleys to the north of K2. The main purpose of this exercise was to link up their mapping work with the 1929 Italian survey of K2 and the Baltoro glacier on the southern flanks of the mountain. They eventually reached the base of its north wall. Shipton was simply overawed:

The cliffs and the ridges of K2 rose out of the glacier in one stupendous sweep to the summit of the mountain 12,000 feet above. The sight was beyond my comprehension ... Sitting alone gazing at it at the cirque forming the head of the K2 glacier was an experience I shall never forget; no mountain scene has impressed me more.

The group then spent the next three weeks exploring the Crevasse Glacier. No one quite knew what lay beyond the unprobed limits of this glacier, but it was conjectured that it would lead to the fabled 'Snow Lake', an extensive basin of snow and ice that had been described by the Workmans and which supposedly formed part of a much vaster ice cap that fed the glaciers from whence they radiated. The Workmans had also claimed they had observed a glacier, which they dubbed the Cornice Glacier, that had no outlet, a contentious assertion on their part to say the least, given the laws of physics. At this point, the group separated. Tilman and a number of Sherpas were sent over the passes at the top of the Crevasse Glacier to settle the issue before making their trek out of the Karakoram and back to civilisation.

When, after crossing various passes, Tilman eventually reached the Snow Lake, he was underwhelmed: *"... this nearly flat basin I should put at about six miles by three; at the most 20 square miles instead of 300. It is a disappointingly small area for such a grandiose name."*

He was more hopeful the Cornice Glacier would defy all the scientific experts however, but inevitably, he was similarly let down. Dryly, he wrote of his experience:

In a drab world it would be refreshing to report the discovery of a glacier flowing uphill, or even of one that did not flow at all. It gives me no pleasure, therefore, to have to affirm that this glacier behaved as others do. ... I ought to have rejoiced, because I can honestly say that to tramp down the Cornice Glacier, hoping at every moment to find an impasse and find none, was as sorry a business as any that has fallen my lot.

By the time Tilman had returned to Askole by way of the Snow Lake and the Cornice Glacier, Shipton and Spender were returning by a different route. They made their way northwards along the Braldu Glacier and Valley to map that uncharted section of country and its river systems, and then recrossed the continental divide using the Shimshal Pass (4,735 m). When they got there, Spender found it, *"hard to believe that we were crossing the main continental watershed, so great was the contrast with the other sections of it which we had visited. It was a wide-open grassy valley near which enormous herds of yaks and sheep were grazing."* Eventually, after three and a half months of filling in chunks of emptiness on the map, they arrived back in civilisation, a sight for sore eyes.

In all, they had surveyed no less than 4,661 square kilometres of terrain. In an autobiography Shipton later wrote:

No experience of mine has been fuller, no undertaking more richly rewarded than those few months among the unknown mountains beyond the crest of the Karakoram. The vast scale of the country, its complete isolation from any source of help or supply demanded all our ingenuity and a wide range of our mountaineering technique. Striving to traverse and understand such a world, and thus to absorb something of its peace and strength was at once our task and our reward.

Endnotes

1 The name *Chogori* has been attributed to the mountain. Derived from two Balti words, meaning 'big mountain', it has been suggested as a local name, but this is probably after K2 had been identified by explorers as a big mountain.

2 Younghusband (1863-1942) would later become the president of the Royal Geographical Society and the first chairman of the Mount Everest Committee which oversaw the first attempts on the peak in the 1920s.

3 In fact, he underspent by just a few pounds.

4 Michael Spender (1906-1945) was brother of the poet, Stephen Spender.

5 John Auden (1903-1991) likewise had a famous poet brother in W.H. Auden.

6 The Sherpas were led by Ang Tharkay (1907-81), described as one of the most outstanding Sherpas of his generation, who had accompanied many well-known Himalayan mountain-climbing expeditions

The 1988 map of Everest was hailed as a masterpiece of Swiss mapmaking and marked the apex of Bradford Washburn's career as a cartographer. One contemporary review predicted that the map would "remain the standard for a generation." Washburn's magnum opus has certainly stood the test of time, and its exquisiteness, practicability and historical value remain unsurpassed today.

Mountaineer turned cartographer

How Bradford Washburn, an erstwhile mountaineer in the American North, came to produce the most exquisite map of Everest yet

Bradford Washburn, one of America's finest mountain explorers, finally took on the challenge of Everest in his seventies. He had first been cast under its spell as a sixteen-year-old boarder at his prep school in Groton, Massachusetts, after listening to a presentation delivered by John Noel, a member of the fateful 1924 British Mount Everest expedition. After an illustrious mountaineering career, followed by over 40 years as director of the Boston Museum of Science, in 1984 Washburn was about to experience the majesty and awe of the world's highest mountain at close hand and from almost every angle. All that was needed was clearance from the Nepalese and Chinese governments and a spell of exceptionally fine weather.

The family that Henry Bradford Washburn Jr. was born into when he came into the world on 7 June 1910 in Cambridge, Massachusetts was of impeccable American stock, with a line of ancestry that could be traced back to the Pilgrim Fathers. He was introduced to the outdoors on family holidays at Squam Lake in New Hampshire, at the northern end of the Appalachian mountain chain. This sparked an insatiable appetite for adventure and he was soon penning accounts of his teenage exploits in the backcountry. His first privately published booklet, an 80-page guide to *The Trails and Peaks of the Presidential Range of the White Mountains*, including trail descriptions, 39 photographs and a neatly drawn pen-and-ink map, attracted the attention of the Appalachian Mountain Club and, still a mere teenager, he was asked to put together the next edition of the club's own guide.

Subsequent trips to Europe allowed the young Bradford to spend summers honing his natural climbing skills in the Alps, and by the time he enrolled at Harvard University at the age of 19, he had already amassed a bucket list of climbs that most outdoor enthusiasts could only hope to achieve in a lifetime, including ascents of Mont Blanc, Monte Rosa and the Matterhorn. Back in America, in the wake of these accomplishments, he was even invited to give public talks about these climbing adventures. When it came to exploration, Harvard held particular attractions: Noel Odell, another veteran of the 1924 British Mount Everest expedition, was visiting lecturer of geology at the university and, like Washburn, he was an active member of the Harvard Mountaineering Club.

It was during his time at Harvard that Washburn was lured by the siren call of the North. By 1925, to all intents and purposes, the peaks of Alaska and the Yukon remained virgin territory to mountaineers and only a few landmark ascents had so far been accomplished. The first of these was the Duke of the Abruzzi's scaling of Mount Saint Elias (5,489 metres) in 1897.[1] The maiden ascent of Mount Blackburn (4,996 metres)[2] in 1912 was notable because it included amongst its party Dora Keen, a pioneering female alpinist of the day. The highest peak on the North American continent at 6,190 metres, Denali (then known as Mount McKinley) was conquered for the first time in 1913 by a party led by Hudson Stuck and Harry Karstens.

For Washburn, the peaks of Alaska and Yukon, with their remoteness and apparent unassailability, represented a similar challenge to those in the Himalayas, on which many of his contemporaries in the United States were now casting their eyes. It was a complete novelty in the early 1930s, but in the Far North, Washburn was starting to open up supply routes in poorly accessible areas by plane, thus averting long and arduous route marches through the wilderness and saving precious time for actual climbing. Air-drops, often in the middle of a glacier, became one of his skills. Careful reconnaissance of the mountain terrain beforehand required strapping himself into the open door of a low-flying plane for aerial photography. It gave him a super keen eye for the topography of the mountains and stood him in great stead for later projects.

In all, Washburn completed thirteen first ascents in the American North, including Mount Crillon (3,879 m) in 1934; Mount Lucania (5,240 m) in 1937; Mount Sanford (4,949 m) and Mount Marcus Baker (4,016 m) in 1938; Mount Bertha (3,110 m) in 1940; and Mount Hayes (4,216 m) in 1941.

In addition to the perils of mountaineering, his hair-raising account of Mount Lucania's[3] conquest in 1937 reflects the significant risks that were involved in these drop-offs in the middle of nowhere. On this occasion, Bob Reeve, one of his

Reconnaissance of the mountain terrain required strapping into the open door of a low-flying plane for aerial photography.

"Anywhere you'll ride, I'll fly." Bob Reeves was Washburn's pilot of choice when it came to remote air-drops.

most entrusted bush pilots – *"Anywhere you'll ride, I'll fly"* – landed Washburn and his climbing partner Bob Bates on an unexpectedly slushy glacier close to the mountain from which any prospect of taking off again soon became almost extinct. Against the odds, after the plane had been stripped of all its excess weight, Reeve finally got airborne after five anxiety-filled days on the glacier. With the chances of a return flight all but scuppered by this episode, Washburn and Bates pushed on to make the first ascent and, in an epic descent and return to civilisation, they faced a 240-kilometre trek through the Yukon wilderness. Without maps, and having abandoned their camera equipment[4] and some of their food to save weight, they then discovered that a cache left by a previous expedition had been plundered by bears, so they had to survive on mushrooms and squirrels. Almost within sight of civilisation, they were then confronted by the heavily swollen Donjek River. This was *"not a mountaineering tale"*, Washburn later recalled, but one of *"a constant struggle with wearying legs and fast diminishing food. We reached the shores of the Donjek after a sixteen-hour battle with damnable moraines, swollen torrents, and packs that seemed constantly to grow heavier. Try as we would, we could not find a single place where we could safely cross the Donjek afoot."* Tired, hungry and without any equipment that might help them negotiate this final obstacle, they at last found a perilously fordable crossing 30 kilometres upstream, before fortuitously running into some native horsepackers heading to their trading post at Burwash Landing.

From the very outset, Washburn's write-ups of his exploits had been catching the eye of the National Geographic Society (NGS). It wasn't just the descriptive accounts of his expeditions which appealed. Despite his young age, Washburn had already developed an expeditionary strategy of *climb, photograph and survey*. He wasn't just evolving as a mountaineer, but as a chronicler too, in every sense of the word. The National Geographic magazine of March 1935 gave an account of his expedition of the previous year to Mount Crillon, together with his photographs and map. That same year, the National Geographic again commissioned his services, this time on an undertaking to map the Saint Elias Mountains range on the US-Canadian border, an as-yet unexplored and uncharted region of the Yukon. After a three-month expedition, in which – amongst other things – he discovered a 65-kilometre-long glacier, the June 1936 edition of the National Geographic published an article entitled *Exploring Yukon's Glacial Stronghold* by Bradford Washburn. Still in his mid-twenties, the NGS offered him a job on the editorial staff of the magazine, which he turned down on the basis that it would place too many restrictions on his hunger to explore.

Instead – with Washburn now aged 28 and employed as a geography instructor at Harvard – he was presented with another opportunity that would irrevocably change the course of his professional and personal life. After a chance meeting on a flight from Boston to Newark in November 1938, he was offered a job as director of the New England Museum of Natural History, later the Boston Museum of Science: its board had an ambition to transform what was a tiresome museum experience stuck in the middle of the last century into an engrossing, entertaining and educational one for visitors. Brad, they felt, was their man. When, after some deliberation, he finally accepted the offer, Washburn took to his new role with positive relish, one that would energise him for the next 40 years. And not only was he smitten by his mission, but by his new secretary too: less than two years into the job, Washburn married Barbara Polk, four years his junior. It was a relationship that would never look back and despite not having acquired any mountaineering experience, Barbara was to become his lifelong climbing partner. Decades later Washburn said, "*Taking the museum job was the second-best decision in my life. Marrying Barbara was the best.*"

Washburn's enthusiasm for exploration was not dimmed by the responsibilities of a career move or the imminent prospect of parenthood, and he was likewise determined his wife should partake in the excitement of his mountaineering escapades. No sooner had they married in April 1940, than Barbara found herself on a train bound for Alaska as part of a team assembled by Washburn for a successful attempt on the as-yet unclimbed peak of Mount Bertha (3,110 metres). One year later, not long after the arrival of their firstborn, Barbara was standing on another virgin summit with her husband, that of Mount Hayes (4,216 metres). Her crowning mountaineering achievement however, was to be the first woman to scale

In 1947, Barbara Washburn was the first woman to summit Denali (Mount McKinley), the highest peak in North America. Here she is pictured with her husband on the top of Mount Bertha in 1940, another first.

Mount McKinley/Denali in 1947. Technically said to be more difficult than Everest, it was only the third ever ascent of the mountain. *"I stood on the roof of North America, the summit of Mount McKinley,"* she wrote in her autobiography, **The Accidental Adventurer**. *"I was very cold. The wind was gusting to thirty miles an hour and the temperature was twenty degrees below zero. And I knew I would be there for a while because our climbing party had survey work to do."*

By 1950, Washburn began to accept that his swashbuckling days as a climber were coming to an end and he threw himself into other more sedate mountain-related projects. In the same way as the Alps had become more easily accessible through the medium of cartography a century earlier, the mountains of Alaska were now being opened up as a playground for mountaineers and he felt that the maps being produced by the US Geological Survey[5] fell short of the mark. In the meantime, his own unrivalled experience exploring Alaska had enabled him to amass so much information on its topography that he deemed it necessary to produce a definitive map of Mount McKinley. Over the years, his many flyovers during which he made use of a bulky Fairfield F-6 camera had provided him with a comprehensive photographic picture of the range, and on the ground, ever since his Crillon expedition in 1934, he had steadily been completing his own triangulation of the mountain range.

In what was to become a longstanding collaboration, Washburn then teamed up with the *Landestopographie der Schweiz*[6] who he believed had the requisite skills to help produce an accurate, detailed and aesthetically pleasing map of Mount McKinley, adapted to the wilderness conditions of the terrain. Faithful representation of rock features, for example, was a particular concern of Washburn's. From conception it took a whole decade before the 1:50,000 map was finally published in 1960. Bearing all the hallmarks of a Swiss topographic map[7], the result is a harmonious balance between beauty and legibility.

In 1965, Washburn set out on another campaign in the Far North. At the time, Canada had recently announced that its highest unclimbed peak, close to the Yukon-Alaska border, would be named in recognition of John F. Kennedy, who had been assassinated two years earlier. As part of the expedition, Washburn not only planned to produce a detailed 1:31,680[8] map of the newly christened Mount Kennedy (4,238 metres), but he also organised an attempt on its summit by a party that included the late president's younger brother Robert.

Washburn coordinated production of the 1:50,000 Mount McKinley (Denali) map. Published in 1960, it represents a harmonious balance between beauty and legibility.

ley, ALASKA

WASHBURN

etts, U.S.A.; The Swiss Foundation for Alpine
d Sciences, and Dr. Alexander Hamilton Rice.

y. CH-3084 Wabern, Switzerland
ndation for Alpine Research, Zürich, Switzerland

211

This time, the sponsors of the expedition were the NGS and Washburn's own Museum of Science in Boston. Whilst not technically difficult, it would be a demanding climb for Senator Kennedy, whose only training had been running up and down the stairs. On 24 March 1965, having slipped down a crevasse on the way up, Robert Kennedy, who three years later was to be himself assassinated in a presidential campaign – was the first to plant a flag on the summit named in honour of his brother. The survey itself that followed was dogged by bad weather, Washburn later explained, "*transforming an otherwise routine expedition into an extremely complex operation.*" Nevertheless, by the winter of that year, a manuscript version had been prepared for Washburn's trusted Swiss cartographers and the map was eventually published in 1968.

His focus shifted south to Arizona for the next great mapping project. On a working visit to the Grand Canyon for the Boston Museum of Science in 1969, Bradford and Barbara Washburn were amazed to discover that no detailed map existed for this geographical wonder. In the National Geographic, Washburn once again found a willing partner, along with the Boston Museum of Science, to co-finance his latest cartographic enterprise. As in previous projects, blanket aerial coverage of the canyon was required to establish the exact lie of the land and then followed up with an on-the-ground survey. Given its labyrinthine topography, with its towering cliffs, vertiginous ravines, multitudinous side-canyons and multi-layered rock stratifications, many of the signal stations were all but unapproachable by foot. To overcome this predicament, Washburn – in inimitable fashion – would have himself, his wife and his equipment transported to otherwise inaccessible pinnacles and rock outcrops by helicopter. Luckily, by the 1970s, he was able to make use of the most advanced surveying equipment of the day, including brand-new laser technology, which enabled mind-bogglingly accurate measurements. In the end it took the best part of seven years to survey a 427-square-kilometre area that made up the Heart of the Grand Canyon. Once the data had been gathered, he needed an equally devoted team of cartographers to craft together a map which met with Washburn's highly rigorous standards. For this, National Geographic could count on the services of their very own Swiss-trained cartographer, Tibor Tóth, who was responsible for the intricate relief depiction. The map was eventually published in the July 1978 issue of the magazine.

In 1980, at the age of seventy, Washburn stepped down from his position as director at the Boston Museum of Science. In retirement[9], the dreams he courted as a young man came back to visit him and his one burning ambition was to create a new, definitive map of Mount Everest. Amongst the best existing maps was a sumptuous 1957 edition produced by Erwin Schneider[10] at a scale of 1:25,000 and published jointly by the German and Austrian Alpine Clubs, and a smaller scale map created by the Royal Geographical Society in 1961 at a scale of 1:100,000. Accurate though they were, both maps still relied heavily on data that had been collected on surveys taken on the northern (Tibetan) side of the mountain in 1921 and 1935 by Oliver Wheeler and Michael Spender respectively, supplemented by on-the-ground photogrammetric measurements on the southern (Nepalese) side which were taken after Nepal opened up to foreigners in 1951. By then, in a reverse scenario, Tibet – now under Chinese control – had closed its borders to outsiders. The Chinese also sent surveying teams to the mountain and completed their first comprehensive survey of the mountain in 1975, albeit mainly of the northern, Tibetan side of Everest. As such, maps of Everest tended to be rather piecemeal affairs, with historical survey data from the northern and southern sides being 'added' together. Washburn aimed to rectify this by using state-of-the-art aerial photography, creating the most complete map of the mountain and its surroundings so far, at the scale of 1:50,000.

Washburn's Grand Canyon map (1978).

A year after retiring, Washburn had a plan in place with his Western backers on board, including long-standing associates – the Boston Museum of Science, the National Geographic Society and the Swiss Federal Office of Topography. But to fly over the mountain, permission first had to be granted by both the Nepalese and Chinese authorities. Whilst fears of point-blank refusal from the Chinese turned out to be ill-founded, negotiations with representatives of the Nepalese king proved somewhat more irksome, and it was only in 1984, after four long years, that Washburn was finally given the green light to enter Nepalese airspace.

Then, shortly before photo surveying was scheduled to take place, came a twist in the tail: on 4 November 1988, Washburn's wife Barbara, days shy of her 70th birthday, was taken seriously ill in her hotel bed in Kathmandu. She was airlifted with her husband, first to Bangkok and then to Boston, where they were told she had cancer. Thankfully, the initial diagnosis proved incorrect and Barbara survived, but with Washburn now half a world away, operations in Nepal had to take place without his hands-on coordination. Nevertheless, with a back-up team in place, on 20 December, a fully equipped and conditioned Learjet took off from Kathmandu airport in perfect weather conditions and, flying at an altitude of some 13,000 metres with an air-speed of 800 kph, returned with 160 razor-sharp negatives.

The data from the aerial photographs was verified against control points measured from NASA's space shuttle, as well as ground measurements recorded by Nepali and Chinese surveyors. Swiss cartographers then got down to the painstaking work of converting all the data into a physical map showing the topography of the Mount Everest range in minute detail, with meticulous attention shown to the structure of the rocks and glaciers. Exhaustive studies were also made into researching the place names and then checked with experts from Nepal and China, which Washburn claimed was the most complex procedure of its production.

A full eight years after its conception, in 1988, the map was published in time to mark the National Geographic Society's centenary and the 150th anniversary of the Swiss Federal Office of Topography. The name Mount Everest appeared on the title of the map, alongside the respective Nepalese and Tibetan names of the mountain, Sagarmatha and Chomolangma. Hailed not only as a masterpiece of Swiss cartography, the map marks the crowning glory of Washburn's career as a cartographer. The review by Michael Ward[11] in the Royal Geographical Society's official journal in 1989 predicted that the map would *"remain the standard for a generation and reflect great credit on all who have been concerned in [its] making."* Washburn's magnum opus has certainly stood the test of time. Its exquisiteness, practicability and historical value remain unsurpassed today and it should have pride of place on any map-lover's shelf.

A 3,000-word essay hardly does justice to the extraordinary life of Henry Bradford Washburn, who died aged 96 in 2007.[12] In the title of his 2009 biography, Dave Roberts refers to Washburn as *"the last of his kind"*, maintaining that *"no adventurer in the twentieth century had a more profound impact on his protégés and imitators, revolutionising the arts of mountaineering and exploration."*

The great Ansel Adams regarded Washburn as an astounding mountain photographer, on a par with Vittorio Sella. In the year prior to his own death in 1984, Adams wrote:

> *Bradford Washburn is one of the very few people who have combined spectacular experience in the wilderness with equally spectacular achievements in the world of civilisation. One never knows what next to expect from this roving genius of mind and mountains, but whatever it is, we know it will be excellent and effective.*

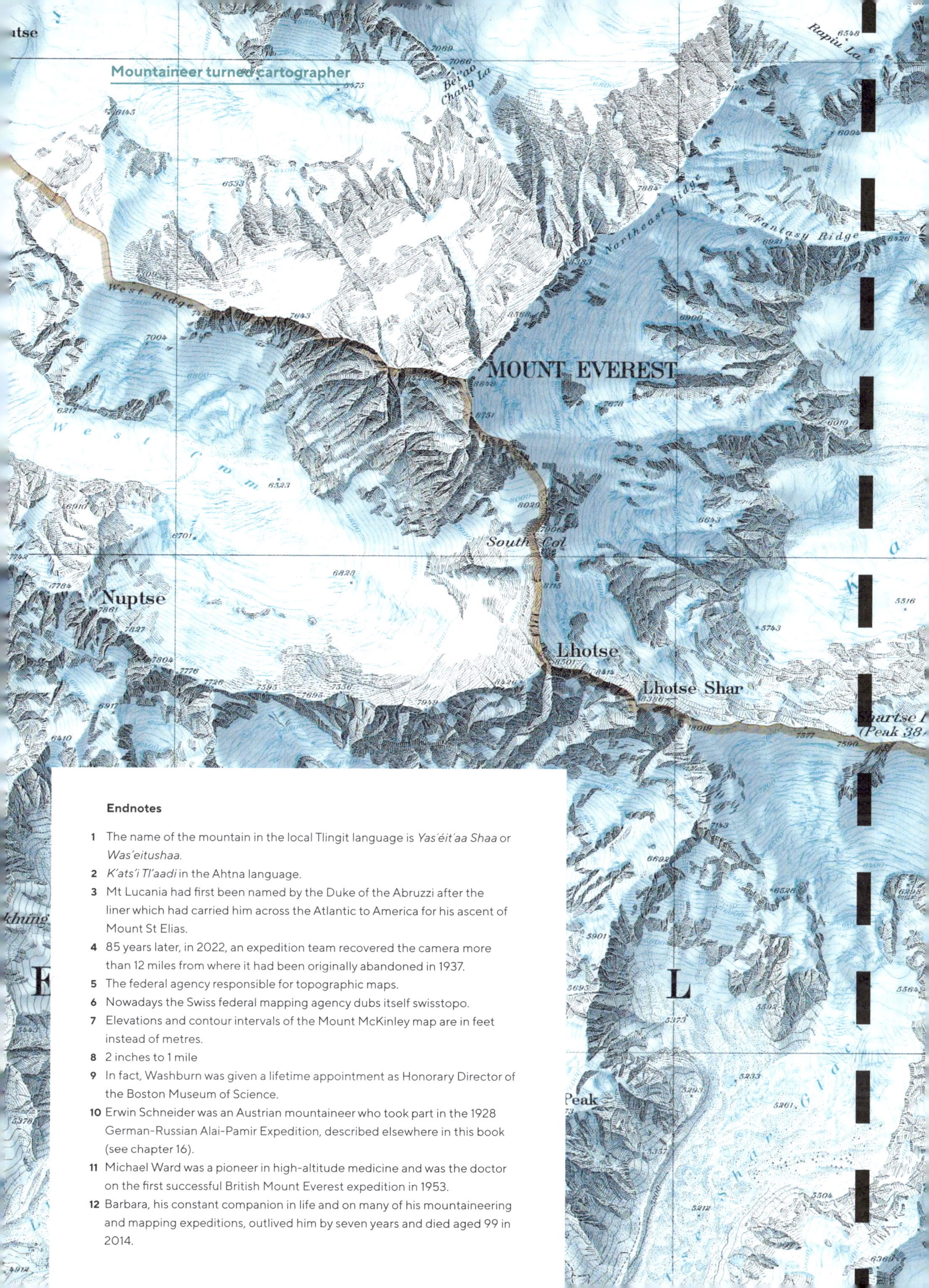

Endnotes

1 The name of the mountain in the local Tlingit language is *Yas'éit'aa Shaa* or *Was'eitushaa*.

2 *K'ats'i Tl'aadi* in the Ahtna language.

3 Mt Lucania had first been named by the Duke of the Abruzzi after the liner which had carried him across the Atlantic to America for his ascent of Mount St Elias.

4 85 years later, in 2022, an expedition team recovered the camera more than 12 miles from where it had been originally abandoned in 1937.

5 The federal agency responsible for topographic maps.

6 Nowadays the Swiss federal mapping agency dubs itself swisstopo.

7 Elevations and contour intervals of the Mount McKinley map are in feet instead of metres.

8 2 inches to 1 mile

9 In fact, Washburn was given a lifetime appointment as Honorary Director of the Boston Museum of Science.

10 Erwin Schneider was an Austrian mountaineer who took part in the 1928 German-Russian Alai-Pamir Expedition, described elsewhere in this book (see chapter 16).

11 Michael Ward was a pioneer in high-altitude medicine and was the doctor on the first successful British Mount Everest expedition in 1953.

12 Barbara, his constant companion in life and on many of his mountaineering and mapping expeditions, outlived him by seven years and died aged 99 in 2014.

Bibliography

General reading

For background information on (historical developments in) mapping, these English-language publications provide a good all-round introduction:

Brotton, Jerry, *A History of the World in Twelve Maps,* (Allen Lane, 2012)

Brotton, Jerry, *Great Maps,* (Dorling Kindersley, 2014)

Garfield, Simon, *On the Map: Why the World Looks the Way it Does,* (Profile Books, 2012)

Mason, Betsy & Miller, Greg, *All Over the Map: A Cartographic Odyssey,* (National Geographic, 2018)

Noble-Wilford, John, *The Mapmakers,* revised edition, (Vintage, 2001)

Storms, Martijn et al., *Maps that Made History: 1000 Years of World History in 100 Old Maps,* (Lannoo, 2022)

The most comprehensive resource on the history of mapping is the *History of Cartography* project. This consists of six volumes, which have been published over a period of almost thirty years by the University of Chicago Press. They are all available online in a PDF-downloadable format: https://press.uchicago.edu/books/HOC/index.html

Volume 1, *Cartography in Prehistoric, Ancient, and Medieval Europe and the Mediterranean,* eds. J. B. Harley & David Woodward (1987)

Volume 2, Book 1, *Cartography in the Traditional Islamic and South Asian Societies,* eds. J. B. Harley & David Woodward (1992)

Volume 2, Book 2, *Cartography in the Traditional East and Southeast Asian Societies,* eds. J. B. Harley & David Woodward (1994)

Volume 2, Book 3, *Cartography in the Traditional African, American, Arctic, Australian, and Pacific Societies,* eds. David Woodward & G. Malcolm Lewis (1998)

Volume 3, *Cartography in the European Renaissance,* ed. David Woodward (2007)

Volume 4, *Cartography in the European Enlightenment,* eds. Matthew Edney & Mary Pedley (2019)

Volume 5, *Cartography in the Nineteenth Century,* ed. Roger J. P. Kain (in preparation)

Volume 6, *Cartography in the Twentieth Century,* ed. Mark Monmonier (2015)

For a history of mountaineering and mountain exploration, with particular respect to the Alps and the Himalayas, these publications provide a good general overview:

The Alpine Club / The Royal Geographical Society, *Mountaineers,* (Dorling Kindersley, 2011)

Douglas, Ed, *Himalaya: A Human History,* (Bodley Head, 2020)

Fleming, Fergus, *Killing Dragons: The Conquest of the Alps,* (Granta, 2000)

Mason, Kenneth, *Abode of Snow; A History of Himalayan Exploration and Mountaineering,* (Dutton, New York, 1955)

Isserman, Maurice & Weaver, Stewart, *Fallen Giants: A History of Himalayan Mountaineering from the Age of Empire to the Age of Extremes,* (Yale University Press, 2008)

Macfarlane, Robert, *Mountains of the Mind: A History of Fascination,* (Granta, 2003)

Underlying principles: surveying and cartographic techniques

Ashworth, Mick, *Why North is Up – Map Conventions and Where They Came From,* (The Bodleian Library, 2019)

Desimini, Jill & Waldheim, Charles et al., *Cartographic Grounds: Projecting the Landscape Imaginary,* (Princeton Architectural Press, 2016)

Imhof, Eduard, *Cartographic Relief Presentation,* (ESRI Press, 2007)

The original English-language version was published in 1982

Kraak, Menno-Jan & Ormeling, Ferjan, *Cartography: visualization of spatial data,* Fourth Edition, (Routledge, 2021)

Parker, Philip, *To the Ends of the Earth: How the greatest maps were made,* (Quarto, 2023)

Pearson, Alastair W., 'Relief Depiction', *History of Cartography: Volume 6,* ed. Mark Monmonier (University of Chicago Press, 2015), 1260-1272

Wallis, Helen M. & Robinson, Arthur H., *Cartographic innovations. An international handbook of mapping terms to 1900,* (Map Collector Publications, 1987)

As well as the publications specified above, documentation consulted in the writing of the specific chapters of the book are listed on the following pages. Many of these publications, particularly journal articles, are available online and/or in a PDF download format.

Bibliography

Per chapter

3. On the itinerary

Carreras, C., de Soto, P., Muñoz, A., 'Land transport in mountainous regions in the Roman Empire: Network analysis in the case of the Alps and Pyrenees', *Journal of Archaeological Science: Reports,* Volume 25, (2019), 280-293

Dilke, O.A.W., 'Maps in the Service of the State: Roman Cartography to the End of the Augustan Era', in *History of Cartography: Volume 1,* ed. by J.B. Harley & D. Woodward, (University of Chicago Press, 1987), 201-211

Snodgrass, A.M., 'The Early History of the Alps', *Alpine Journal,* (1993), 213-222

Swiss-style Part I

Aliprandi, Laura & Giorgio / Pomella, Massimo, *Le Grandi Alpi nella Cartografia dei Secoli Passati 1482-1865,* (Priuli & Verlucca, 1974)

Balmer, Heinz, 'Konrad Türst und seine Karte der Schweiz', *Gesnerus: Swiss Journal of the History of Medicine and Sciences,* (1972), 79-102

Gattlen, Anton, 'Die älteste Walliserkarte', *Cartographica Helvetica,* (1992), 31-40

Hubert, Samuel & Roth, Simon, *Le Valais à la carte: 1'000 déclinaisons d'une collection,* (Editions Monographic, Sierre, 2023)

Karrow Jr, Robert W., *Mapmakers of the Sixteenth Century and Their Maps,* (Speculum Orbis Press, Chicago, 1993)

Koller-Weiss, Katharina, 'Aegidius Tschudis grosse Manuskriptkarte des schweizerischen Raums und der angrenzenden Gebiete, um 1565', *Cartographica Helvetica,* (2005), 3-16

Meurer, Peter H., 'Cartography in the German Lands, 1450 -1650', in *History of Cartography: Volume 3,* ed. D. Woodward, (University of Chicago Press, 2007), 1172-1245

5. The Cassinis and the birth of modern mapping

Pelletier, Monique, 'Cartes de France', *History of Cartography: Volume 4,* eds. Matthew Edney & Mary Pedley, (University of Chicago Press, 2019), 239-244

Pelletier, Monique, 'Cassini Family', *History of Cartography: Volume 4,* eds. Matthew Edney & Mary Pedley, (University of Chicago Press, 2019), 251-258

Pelletier, Monique, 'Jean Picard', *History of Cartography: Volume 4,* eds. Matthew Edney & Mary Pedley, (University of Chicago Press, 2019), 1095-1097

Rand Hoare, Michael, 'Expeditions to Lapland and Peru', *History of Cartography: Volume 4,* eds. Matthew Edney & Mary Pedley, (University of Chicago Press, 2019), 727-731

Swiss-style Part II

Bürgi, Andreas, 'Der Blick auf die Alpen: Franz Ludwig Pfyffers Relief der Urschweiz (1762 bis 1786)', *Cartographica Helvetica,* (1998), 3-9

Cavelti Hammer, Madlena, 'Herstellung und Auswirkungen des Reliefs der Urschweiz von Franz Ludwig Pfyffer', *Cartographica Helvetica,* (1998), 11-18

Wyder, Samuel / Feldmann, Hans-Uli, 'Die Karten der Schweiz (1620-1657) von Hans Conrad Gyger', *Cartographica Helvetica,* (2011), 3-18

6. Oranges and lemons

Crane, Nicholas, *Latitude: the astonishing journey to discover the shape of the earth,* (Michael Joseph, 2021)

Solarić, Miljenko & Solarić, Nikola, 'The French Scientific Surveying Expedition to Peru', *Kartografija i Geoinformacije* Vol. 13, No. 21, (2014), 4-19

Whitaker, Robert, *The Mapmaker's Wife,* (Basic Books, New York, 2004)

7. William Roy's Great Map

Ang, Tim & Pollard, Michael, *Walking the Scottish Highlands: General Wade's Military Roads,* (André Deutsch, 1984)

Fleet, Christopher & Kowal, Kimberly C., 'Roy Military Survey map of Scotland 1747-1755: mosaicing, geo-referencing, and web delivery', *e-Perimetron,* Vol. 2, No. 4, (2007), 194-208

Hewett, Rachel, *Map of a Nation: a biography of the Ordnance Survey,* (Granta Books, 2010)

Mallett, R.J., *The Military Survey of Scotland 1747-1755: An Analysis Utilising the Dual Concepts of Map Form and Content,* PhD diss., (University of Aberdeen, 1987)

Owen, Tim & Pilbeam, Elaine, *Ordnance Survey: Map Makers to Britain Since 1791,* (Ordnance Survey & HMSO, 1992)

8. The Stargazer

Andra-Warner, Elle, *David Thompson: A Life of Adventure and Discovery,* (Heritage House, 2010)

Jenish, D'Arcy, *Epic Wanderer: David Thompson & the Mapping of the Canadian West,* (University of Nebraska Press, 2003)

Shoalts, Adam, *A History of Canada in Ten Maps: Epic Stories of Charting a Mysterious Land,* (Penguin Random House, 2017)

9. The unlikely mountaineer

Arvers, Paul, 'Le Capitaine Durand et le Pelvoux', *Annuaire du Club alpin français,* (1887), 3-12

Beraldi, Henri, 'Le Capitaine Durand, 1787 - 1835', *La Montagne,* No. 7, (1911), 1-19 & 65-93

Guilhot, Nicolas, *Histoire d'une parenthèse cartographique : Les Alpes du Nord dans la cartographie topographique française aux 19e et 20e siècles,* Doctoral thesis, Université Lyon, (2005)

Mille, Jacques, Barféty, Jean-Marc & Tailland, Michel, *Le massif des Écrins, Histoire d'une cartografie de l'Antiquité à l'aube du XXe siècle,* (Éd. Fournel, 2019)

Whymper, Edward, *Scrambles amongst the Alps,* (1870)

10. Napoleon's triangles

Geilen, Quirin, 'Entstehung der Tranchot-Müfflingschen Kartenaufnahme und des rheinischen Kadasters', *Mit Wasser und Dampf,* Ed. Fehl, G., Kaspari-Küffen, D., Meyer, L-H, (Meyer & Meyer, 1991), 212-213

Schmidt, R., 'Die Kartenaufnahme der Rheinlande durch Tranchot und v. Müffling 1801 – 1828', *Gesellschaft für Rheinische Geschichtskunde XII* (1973)

11. The Great Arc

Keay, John, *The Great Arc: The Dramatic Tale of How India was Mapped and Everest was Named,* (HarperCollins, 2000)

Philimore, R.H., *Historical Records of the Survey of India,* Volumes I-V, (Dehra Dun, 1950-68)

12. The steppe-counters

Dean, Riaz, *Mapping the Great Game: Explorers, Spies and Maps in 19th-Century Asia,* (Casemate, 2019)

Hopkirk, Peter, *The Great Game: On Secret Service in High Asia,* (John Murray, 1990)

Hopkirk, Peter, *Trespassers on the Roof of the World: The Race for Lhasa,* (John Murray, 1982)

Keay, John, *When Men & Mountains Meet: Explorers of the Western Himalayas, 1820-75,* (John Murray, 1977)

Montgomerie, T.G & Pundit, 'Report of a Route-Survey Made by Pundit, from Nepal to Lhasa, and Thence Through the Upper Valley of the Brahmaputra to Its Source', *The Journal of the Royal Geographical Society of London,* Vol. 38, (1868), 129-219

Moorhead, Catherine, *The K2 Man (and his Molluscs): The Extraordinary Life of Haversham Godwin-Austen,* (Neil Wilson Publishing, 2013)

Waller, Derek, *The Pundits: British Exploration of Tibet and Central Asia,* (The University Press of Kentucky, 1990)

Ward, Michael, 'The Survey of India and the Pundits: the Secret Exploration of the Himalaya and Central Asia', *The Alpine Journal,* (1998), 59-79

Alpine Clubs

Brunner, Kurt & Welsch, Walter, 'High-mountain cartography of the German and the Austrian Alpine Clubs', *ISPRS Journal of Photogrammetry and Remote Sensing,* Volume 57, (2002), 126-133

Kostka, Robert & Moser, Gerhardt, *Die Kartografie im Alpenverein: an der Schwelle zum 21. Jahrhundert,* Oesterreichische Alpenverein & Deutsche Alpenverein, (Weishaupt Verlag, 2014)

Gurtner, Martin, 'The Swiss Alpine Club SAC and Its Role in national Mapping', *Proceedings of the 6th ICA Mountain Cartography Workshop* ('Mountain Mapping and Visualisation'), Lenk, Switzerland, (2008), 81-84

13. Twenty-one summers

Helbronner, Paul, *La Description Géométrique des Alpes Françaises,*

Tome X : *Massifs du Champsaur, du Devoluy, des Écrins et du Pelvoux - campagnes 1904-05-06-18-21-22,* (Gauthier-Villars, Paris, 1935)

(The volume on Les Écrins is the last in a set of ten volumes)

Helbronner, Paul, 'Quatre mois de triangulation géodésique dans le massif Pelvoux-Écrins', *La Montagne,* Vol. 2, No. 1, (Paris, January 1906)

Leon, Daniel, *Les Alpes d'Helbronner: mesure et démesure,* (Éditions Glénat, Grenoble, 2015)

Palmieri, Joseph, 'Paul Helbronner (1871-1938) et la Géodésie de la Corse', *La société des sciences historiques et naturelles de la Corse et son temps,* (2010), 229-277

14. To the moon and back

Adams, Ansel et al., Summit: Vittorio Sella, mountaineer & photographer 1879-1909, (Aperture, 1999)

Bere, R.M., 'The Exploration of the Ruwenzori', Alpine Journal, Vol. 58, (1952), 483-495

Filippi, Filippo de, Ruwenzori: an account of the expedition of HRH Prince Luigi Amadeo of Savoy, Duke of the Abruzzi, (Dutton & Company, New York, 1908)

Jeal, Tim, Explorers of the Nile: The Triumph and Tragedy of a Great Victorian Adventure, (Faber & Faber, London, 2011)

Montovani, Roberto, *The Ruwenzori discovery: Luigi Amedeo di Savoia duca degli Abruzzi*, (Museo Nazionale della Montagna, CAI, Turin, 1996)

Pennacini, Cecilia, ed., I Popoli della Luna – The People of the Moon, Ruwenzori 1906-2006, (Museo Nazionale della Montagna, CAI, Turin 2006)

Tenderini, Mirella & Shandrick, Michael, The Duke of the Abruzzi: An Explorer's Life, (Mountaineers Books, Seattle, 1997)

Tilman, H.W., Snow on the Equator, (Bell & Sons, 1937)

Swiss-style Part III

Cavelti Hammer, Madlena, 'Zum 100. Todesjahr von Xaver Imfeld (1853-1909): Ingenieur und Alpenkartograph', *Cartographica Helvetica*, (2009), 3-10

Cavelti Hammer, Madlena et al., *Xaver Imfeld, 1853-1909: Meister der Alpentopografie*, (AH Druck, 2006)

Feldmann, Hans-Uli et al., 'Guillaume-Henri Dufour : Vermessung und Kartierung der Schweiz', *Cartographica Helvetica*, Heft 54, (2017)

Frei, Philippe, & Frey, Fellix et al., *Die Schweiz Auf Dem Messtisch: 175 Jahre Dufourkarte*, (Schwabe Verlagsgruppe AG (2020)

15. Everest finally on the map

Davis, Wade, *Into the Silence: The Great War, Mallory and the Conquest of Everest*, (Vintage, 2012)

Howard-Bury, Charles Kenneth et al. *Mount Everest, the reconnaissance, 1921*, ('Appendix I: The Survey' by H.T. Morshead and 'Appendix II: The Photographic Survey' by E.O. Wheeler), (1922)

Keay, John, *The Great Arc: The Dramatic Tale of How India was Mapped and Everest was Named*, (HarperCollins, 2000)

Wheeler, E.O., 'The Mount Everest Expedition, 1921', *Canadian Alpine Journal*, Vol. 13. (1923), 1-25.

16. The slithering giant

Borchers, Philipp & Wien, Karl (with contributions by Allwein, Eugen & Schneider, Erwin), 'Bergfahrten im Pamir', *Zeitschrift des Deutschen und Österreichischen Alpenvereins*, Band 60, (1929), 64-142

Borchers, Philipp, 'Im Dienste der Erdkunde, Ein Erlebnis auf der Alai-Pamir-Expedition', *Jahresbericht des Akademisches Alpenvereins*, (Berlin 1927-28), 27-34

Finsterwalder, Richard, 'Das Expeditionsgebiet im Pamir: Zu den Bild und Kartenbeilagen der Alai-Pamir-Expedition 1928', *Zeitschrift des Deutschen und Österreichischen Alpenvereins*, Band 60, (1929), 143-160

Rickmers, Willi R., 'Die Alai-Pamir-Expedition 1928', *Zeitschrift des Deutschen und Österreichischen Alpenvereins*, Band 60, (1929), 59-63

Rickmers, Willi R., 'On the Fringe of the Pamirs', *Alpine Journal*, Vol. 41, (1929), 156-160

Rickmers, Willi R., 'The Alai-Pamirs in 1913 and 1928', *The Geographical Journal*, Vol. 74, No. 3 (Sep 1929), 209-228

Schneider, Erwin, 'Pik Lenin 7130 m', *Jahresbericht des Akademisches Alpenvereins*, (Berlin 1927-28), 35-41

Swiss-style – Part IV

Cavelti Hammer, Madlena et al., 'Farbe, Licht und Schatten: die Entwicklung der Reliefkartographie seit 1660', *Cartographica Helvetica*, Sonderheft, (1997)

Imhof, Eduard, *Cartographic Relief Presentation*, ESRI Press (2007).

The original English-language version was published in 1982.

Feldmann, Hans-Uli, 'Charles Jacot-Guillarmod: One of the Pioneers of the Swiss Style of Rock Representation' for the *Proceedings of the 6th ICA Mountain Cartography Workshop* ('Mountain Mapping and Visualisation'), Lenk, Switzerland, (2008), 41-47

Originally published in German for *Cartographica Helvetica*, (2005): 'Der Topograph Charles Jacot-Guillarmod (1868–1925): Einer der Pioniere der Schweizer Felsdarstellung'

See also:

The making of the 1960 Mount McKinley / Denali map: https://www.swisstopo.admin.ch/en/mckinley-map-20211006

The making of the 1988 Mount Everest map:https://www.swisstopo.admin.ch/en/everest-map-20171211

17. Off the map

Baker, Deborah, *The Last Englishmen: Love, War, and the End of Empire*, (Graywolf, 2018)

Filippi, Filippo de, *Karakoram and Western Himalaya, 1909: An Account of the Expedition of H.R.H. Prince Luigi Amedeo of Savoy, Duke of Abruzzi*, (Dutton, New York, 1912)

Jacot-Guillarmod, Jules, *Six Mois dans l'Himalaya, le Karakorum et l'Hin-du-Kush: Voyages et Explorations aux plus Hautes Montagnes du Monde*, (Sandoz, 1904)

Perrin, Jim, *Shipton and Tilman: The Great Decade of Himalayan Exploration*, (Hutchinson, 2013)

Shipton, Eric, 'The Shaksgam Expedition 1937', *The Alpine Journal*, Vol. 50, (May 1938), 34-59

Or:

Shipton, Eric, Spender, Michael, & Auden, John, 'The Shaksgam Expedition 1937', *The Geographical Journal*, Vol. 91, No. 4, (April 1938), 313-336

Shipton, Eric, *Blank on the Map: Pioneering Exploration in the Shaksgam Valley and Karakoram Mountains*, Vertebrate Publishing (2019).

First published in London by Hodder & Stoughton in 1938

Spender, Michael, 'The Shaksgam Expedition 1937', *The Himalayan Journal*, Vol. 10, (1938)

Workman, Fanny Bullock & Workman, William Hunter, *In the ice world of Himálaya, among the peaks and passes of Ladakh, Nubra, Suru, and Baltistan*, (Fisher Unwin, 1900)

18. Mountaineer turned cartographer

Roberts, David, *The Last of His Kind: The Life and Adventures of Bradford Washburn, America's Boldest Mountaineer*, (HarperCollins, 2009)

Scott, Chic, *Pushing the Limits: The Story of Canadian Mountaineering*, (Rocky Mountain Books, 2000)

Ward, Michael, 'Sagarmatha—Mount Everest—Chomolungma. Map by the National Geographic Society', *Geographical Journal*, 155, (1989), 433-435

Washburn, Bradford, 'The Ascent of Mt Lucania', *American Alpine Journal*, (1938), 95-108

Washburn, Bradford, 'First Ascent of Mount Kennedy and Second Ascent of Mount Hubbard', *American Alpine Journal*, (1966), pp 93-97

Washburn, Bradford, 'Mapping Mount Everest', *Bulletin of the American Academy of Arts and Sciences*, Vol. 42, No. 7, (1989), 29-44

Washburn, Bradford & Roberts, David, *Mount McKinley: The Conquest of Denali*, (Abrams, New York, 1991)

Washburn, Bradford & Smith, Donald, *On high: the adventures of legendary mountaineer, photographer, and scientist*, (National Geographic, Washington, 2002)

Picture and map credits

We have done our utmost to trace sources and right holders of images shown. If, despite this, images are shown of which you are (co-)entitled and for the use of which you have not granted permission, we kindly ask you to contact us.

Alamy (16, 18, 22, 22, 36, 85, 118)

Alaska State Library Map Collection (210-211)

Alaska State Library, Leonard Delano Photo Collection (207)

Archiv des Deutschen Alpenvereins, München (10, 25, 135, 136, 178-179, 181, 183)

Archivo General de la Nación Colombia (70-71, 73)

Andrew Davies (45, 46, 107, 113, 222)

Biblioteca Nazionale del Club Alpino Italiano (2, 138, 148-149, 155, 161, 167, 176-177, 188-189, 194-195, 198, 199)

Bibliothèque Dauphinoise (Cover, 4, 108-109)

Bibliothèque Nationale de France (17, 52, 102, 104, 144)

Bibliothèque nationale de France, département Arsenal (143, 146-147)

Bibliothèque nationale de France, département Société de Géographie (130)

Bibliothèque Nationale de France (public domain) (60, 63)

Bundesamt für Landestopografie swiss-topo (25, 27, 28, 28, 165)

David Rumsey Historical Map Collection (75, 26, 193)

David Thompson fonds / Archives of Ontario (88, 97)

ETH-Bibliothek Zürich (50, 166-167)

Fondazione Sella, Biella (156, 157, 159, 162, 201)

Fondo Antiguo de la Biblioteca de la Universidad de Sevilla (74)

Géoportail IGN (64-65, 89-99)

Gesellschaft für Rheinische Geschichtskunde / R.Schmidt (115, 115)

Google Maps (180)

John Brown (2022). Pundit Maps. Zenodo (122, 131)

Kadaster / PDOK (27, 116)

Leiden University Libraries (151)

Library and Archives Canada / Mountain Legacy Project (24)

Library and Archives Canada / Charles William Jefferys fonds (91)

Library of Congress (133)

Musée Dauphinois Grenoble – Département de l'Isère (146, 140)

Museum of Science, Boston (206, 209)

National Geographic Maps (213)

National Library of Scotland (27, 76-77, 87, 80-81, 86)

Princeton University Library (129)

Public domain (39, 42,48,49, 101)

Roelant Meijer (8-9, 20, 23, 153)

Royal Geographical Society (168-169, 172, 173, 174-175, 175, 197, 202)

Science Museum London (23)

Shutterstock (47)

St. Gallen, Stiftsbibliothek (57)

Swiss Foundation for Alpine Research (204, 215)

Thomas Fisher Rare Book Library / University of Toronto (93-95)

Toon Hezemans (35)

Universitätsbibliothek Basel (58)

Universitätsbibliothek Bern (68-69)

University of Texas Libraries, Austin (126-127)

VU Beeldbank cartografisch materiaal (110-111, 116)

Wikimedia Commons (6, 18, 19, 21, 40-41, 82, 103, 117, 191, 120, 121, 68)

Wikimedia Commons / Musée Dauphinois (32)

Yule Collection: Album of Portraits (125)

Zentralbibliothek Zürich (55, 59, 67, 164)

Zentralbibliothek Zürich / David Rumsey Map Collection (67)

Andrew Davies

Andrew Davies was born in Stretford, Manchester in 1957, a stone's throw away from Old Trafford football and cricket grounds, both of which became an integral part of his growing up in neighbouring Chorlton-cum-Hardy. He developed a passion for maps, the outdoors and languages at an early age and went on to study Geography at Newcastle University, graduating in 1979. After several years working in publicity as a copywriter, he moved to Heerlen in 1984, swapping the moorlands of Northern England for the hills of Zuid-Limburg in the Netherlands. In the decade that followed, he was employed on various European projects in the field of disability, before becoming a full-time self-employed translator in 1995. Over the last 30 years he has also acquired degrees in photography and English language teaching and between 2003 and 2018 ran Heerlen Cricket Club which played at a competitive level in both the Dutch and Belgian leagues. Since 2010, Andrew has been an active member of the Dutch Mountain Film Festival team, likewise based in Heerlen. He organised the Mapping the Mountains exhibition for the 2019 festival and following its success was asked to write this book on mountain cartography.

Alastair Bonnett

Alastair Bonnett is Professor of Social Geography at Newcastle University. His travel and academic books have been translated into 19 languages and include *Off the Map, Beyond the Map, The Age of Islands, 40 Maps that Will Change How You See the World* and *Extraordinary Islands: An atlas of unlikely places.* He has also contributed to history and current affairs magazines on a wide variety of topics, such as world population and radical nostalgia. Alastair lives in Newcastle-upon-Tyne.

Acknowledgements

Maps and mountains have been a constant source of cheer in my life, so writing about them requires little motivation. Nevertheless, the help and encouragement of others have, at times, proved indispensable in the making of this project. It would never have happened if Thijs Horbach and Toon Hezemans of the *Dutch Mountain Film Festival* (DMFF) hadn't pushed me headlong down a rabbit hole of maps and mountains which has held me spellbound for the last six years. Toon especially has served as my go-to sounding board for the book throughout.

I can't speak highly enough of the staff at the *Biblioteca Nazionale del Club Alpino Italiano* in Turin. Their digitised database of historical maps was a useful starting point and on my many visits, Alessandra Ravelli (head of documentation) and her colleagues were always on hand to produce the research material I needed and we even spent one New Year's Eve scanning maps in the library. As for *Fondazione Sella* in Biella, which is dedicated to preserving the photographic heritage of Vittorio Sella, Andrea Pivotto (custodian of Sella's archives) welcomed me with open arms on my two visits there. The Fondazione has generously supported the publication with photographs taken by the grand master.

Martijn Storms, curator of maps and atlases at the *University of Leiden*, who I met on a number of occasions, was also kind enough to review several of the texts. His advice on the history of cartography was crucial in shaping some of the earlier chapters. I am also indebted to Samuel Hubert (currently research associate for the *Musée du Vin* in Sierre) and Felix Frey (historian at *swisstopo*) for their helpful comments on Swiss cartography. Likewise, my thanks go to Paula Williams (Curator of Map, Mountaineering and Polar Collections at the *National Library of Scotland*) for sharing her insights on William Roy's map.

Sourcing the maps and illustrations for the book was a particular concern. I needn't have worried: my DMFF colleague Karin Eken went about this task with great gusto, sourcing all the required material. Karin has also been responsible for project planning and generally keeping me on my toes. Journalist and fellow DMFF team member, Ernst Arbouw gave valuable support and feedback on style and content. I am grateful to mountaineering guru, Bob A. Schelfhout Aubertijn for fact-checking and to Allison Klein, a good friend and translation colleague of many years, for proofreading and editing the final copy.

Roelant Meijer of *Tegenwind* has carried out many award-winning design projects, amongst others for the DMFF, and also produces original works about his own peregrinations on foot. He has conjured up a wonderful design for Mapping the Mountains and has likewise been a pleasure to work with.

I am delighted Alastair Bonnett agreed to write the foreword. Apart from being an acclaimed author on all things geographical, Alastair is professor of geography at my alma mater, Newcastle University, just one of the many stops on my lifelong journey in the world of maps.

Over the course of my work on this book, there have been many chance meetings and conversations with inspiring people, too many to mention individually. Friends and family - including my late mum and dad who instilled in me a love of the hills - have also given me their unstinting support. All of you have played your part in the book's creation.

Finally, Francesco Giani provided transport, accommodation and company for some memorable on-the ground research in Les Écrins and the Cottian Alps.
I could, of course, not fail to thank my partner Sue, who likewise accompanied me - sometimes unbeknown - on several of my research trips and over time has grown to understand my quirkiness for the engrossing story of maps and mountains.

Andrew Davies

Credits

Concept: Toon Hezemans, Dutch Mountain Film Festival
Author: Andrew Davies
Foreword: Alastair Bonnett©
Production: Karin Eken
Consultant editor: Ernst Arbouw
Text editing: Allison Klein, ABK Translations
Book design & layout: Roelant Meijer, Tegenwind
Printing: Booxs

ISBN 978-90-832223-5-6

Maps on cover and opening pages
Cover and p.4 - Carte Topographique du Massif Du Mont Pelvoux, 1874
p.2 - Spedizione Geografica Italiana nel Caracorum, 1929
p.6 - Sketch Map of Mount Everest from Surveys by Major Wheeler in 1921 with additions by the Expedition of 1924
p.10 - Karte der Allgaeuer & Lechtaler Alpen, 1906

Dutch Mountain Film Festival

www.mappingthemountains.eu